Landforms of the Earth

Francisco Gutiérrez • Mateo Gutiérrez

Landforms of the Earth

An Illustrated Guide

Springer

Francisco Gutiérrez
Earth Sciences Department
University of Zaragoza
Zaragoza, Spain

Mateo Gutiérrez
Earth Sciences Department
University of Zaragoza
Zaragoza, Spain

ISBN 978-3-319-26945-0 ISBN 978-3-319-26947-4 (eBook)
DOI 10.1007/978-3-319-26947-4

Library of Congress Control Number: 2016930684

Printed on acid-free paper

This Springer imprint is published by Springer Nature
The registered company is Springer International Publishing AG Switzerland

Nieves, Marina, Isabel and Francho

Preface

This book has been produced by two university lecturers with long experience in teaching geomorphology and other related subjects. Students enrolled in courses dealing with earth-surface processes and landforms (geomorphology, physical geography, geology, geohazards, environmental sciences) have access to excellent geomorphology textbooks with superb descriptions. However, only some of the landforms explained in those publications are illustrated with images. This hinders the comprehension of explanations on the landforms and genetic processes and the students' capability to correctly diagnose geomorphic features, both in the field and using remote-sensed data. We feel that visualization of landforms is essential for learning geomorphology and stimulating the interest on this field-based subject; a picture is worth a thousand words.

Nowadays, researchers dealing with geomorphological issues are overwhelmed by an enormous variety of rapidly evolving techniques. Some of them, like dating methods, remote-sensing, or geographical information systems, have greatly contributed to the recent renewed interest in geomorphology. However, the means should not set aside the aims. Basic knowledge is essential to conduct compelling geomorphological investigations, regardless of the sophistication of the methods used. We may use the best available remote-sensed data, but we only see what we know. Remote-sensed data may provide a great deal of information on the landforms, but we should feel forced to examine them directly in the field. The new techniques can complement the classical field-based approaches, but cannot replace most of them.

We hope this photographic guide will constitute an useful educational resource for students, researchers, and practitioners. It may also satisfy the curiosity of people keen on nature, travellers and landscape photographers, who want to learn about the names and origin of the landforms they encounter in their trips. Commonly, stunning geomorphological features constitute the main highlight of protected areas (National Parks, UNESCO World Heritage Sites) and outstanding tourist attractions.

The book illustrates with more than 360 photographs and satellite images from all over the world the main landforms of the Earth's continental surface. The good quality colour images have been thoroughly selected, showing representative and dramatic examples. Each landform or group of geomorphic features is described with concise, direct, and easy-to-read texts. The structure of the book follows classical subdivisions of geomorphology, with five chapters dealing with structural geomorphology (structural, tectonic, volcanic, karst, and granite landforms), four devoted to azonal geomorphic systems and processes (weathering, slope movements, fluvial, and coastal landforms), and three chapters covering the three main morphoclimatic environments (glacial, periglacial, and desert-aeolian landforms). Readers can easily find more lengthy and in-depth explanations in the general and specific geomorphology textbooks recommended at the end of the book.

Zaragoza, Spain

Francisco Gutiérrez
Mateo Gutiérrez

Acknowledgements

We, the authors, are very grateful to the colleagues who have reviewed the different sections of the book providing insightful comments: Anthony Cooper and Kenneth Johnson (introductions), Clifford Ollier (structural and weathering landforms), James McCalpin (tectonic landforms), Juan Carlos Carracedo (volcanic landforms), Ira Sasowsky (karst landforms), Piotr Migoń (granite landforms), Mauro Soldati (slope movements), Adrian Harvey (fluvial landforms), Alan Trenhaile (coastal landforms), Xavier Mir (glacial landforms), Gonçalo Vieira (periglacial landforms), and Andrew Goudie (desert and aeolian landforms).

We would also like to thank those who have kindly given permission to reproduce their images: Franck Audemard, Philippe Audra, Javier Baztán, Nicolas Benoist, Jean-Yves Bigot, Janina Borysiak, Ferrán Cardona, Juan Carlos Carracedo, Rosa Carrasco, Márcio Décio, Damien Closson, Anthony Cooper, Donald Crain, Beth van Dusen, Germán Flor, José María García-Ruiz, Ángel Ginés, Juan Carlos Gordillo, Paula James, Cezary Kabała, Alfredo Lainez, Tim Lawnicki, James McCalpin, Vincent Matthews, Tony Merino, Piotr Migoń, Xavi Mir, John Mylorie, Mohammad Hasan Nazari, José Ortega, Arthur Palmer, Arturo Polo-Ena, Tom Rayner, Carles Roqué, Ira Sasowsky, Matti Seppala, Servicio de Cartografía del Principado de Asturias, Shahram Sherkati, Keith Smith, Scott Weatherson, Javier Ventosa, Virtualgeo, Jon White, Paul Williams, and Zbigniew Zwoliński.

Contents

1 Introduction

Geomorphology literally means the study of the Earth's landforms. It is defined by the International Association of Geomorphologists (IAG; www.geomorph.org) as the interdisciplinary and systematic study of landforms and their landscapes, as well as the Earth surface processes that create them. Geomorphologists investigate landforms and landscapes addressing multiple aspects, such as their genesis, morphometry, chronology and past and future evolution. They also investigate Earth surface processes not only to understand their morphogenetic role, but also to gain a scientific basis for assessing and managing a great deal of environmental problems associated with them, notably geohazards (e.g. landslides, floods, subsidence, earthquakes).

The concept of geomorphology has rapidly evolved over its short history. During its initial stages, the pioneers of the discipline were mainly interested on the reconstruction of the long-term evolution of present-day landscapes (historical geomorphology). That was a period during which geoscientists were largely focused on obtaining information about the past. This retrospective approach was commonly based on the application of the uniformitarianism principle: "the present is the key to the past". In a subsequent phase, qualitative descriptions were replaced by quantitative studies on both processes and landforms, largely thanks to the development of new technologies. Measuring the dimensions of landforms (morphometry), analysing their spatial distribution and frequency and quantifying rates of processes and the associated morphological changes are some of the main targets of quantitative geomorphology. Both historical and quantitative investigations have benefited by the advent of new dating techniques that allow determining the age of landforms and deposits and calculating temporal frequencies and rates; no dates no rates. Recently, there has been a significant trend to applying geomorphology as a tool for assessing and mitigating environmental problems like hazards and impacts (applied or environmental geomorphology). Nowadays, a major challenge is the development of reliable predictions on the future behaviour of Earth surface processes and landforms. This prospective approach is usually developed by using the present and the past as the key to the future, through a reverse uniformitarianism concept. Modern geomorphologists are more worried about the surface processes that operate today and those that will have consequences tomorrow, because they affect our society. This concern is largely motivated by three factors: (1) Damage related to various hazardous geomorphological processes (e.g. floods, landslides) is increasing exponentially. (2) Humans are currently one of the main geomorphic agents by causing direct changes on the land surface or by producing disturbances that significantly change the dynamics of geomorphic systems. From 1900 AD to 2000 AD, the world population increased fourfold to almost six billion people, and economic activity increased 40-fold. In fact, a new geological epoch called the Anthropocene has been proposed, covering the period during which human activities have had a profound imprint on the Earth's landscapes and ecosystems. (3) There is a need to forecast the numerous impacts of the anthropogenic global warming on geomorphic systems, some of which will have significant detrimental effects for our society (e.g. sea-level rise). The current boom of geomorphology is most probably due to its social relevance.

Landscapes are the result of the competing effects of endogenous processes, like tectonic activity and volcanism that create topography, and exogenous processes that tend to erode upland regions and accumulate sediments in lowlands and subsiding areas. The distribution and configuration of the large morphostructural units of the planet are largely governed by plate tectonics (megageomorphology). A small fraction of the energy involved in the horizontal translation of the lithospheric plates is converted into the work necessary to uplift rocks against gravity. Solar energy is the main engine for Earth surface processes. It evaporates water and raises moist air in the atmosphere. The vapour condenses at high elevation and falls as precipitation. The potential energy of the meteoric water is transformed into kinetic energy and expended by doing geomorphic work, including erosion and transport of sediments.

Surface processes, controlled by a number of extrinsic and intrinsic factors, may create and modify landforms.

F. Gutiérrez, M. Gutiérrez, *Landforms of the Earth*, DOI 10.1007/978-3-319-26947-4_1

Some changes in the landforms are related to processes or stimuli that exceed a critical threshold (e.g. landslide triggered by earthquake shaking). Those adjustments may be immediate or can be achieved over a certain period, called the response time (e.g. isostatic rebound related to deglaciation). The magnitude and frequency patterns of morphogenetic processes are highly variable (gradualism versus catastrophism). Some processes produce continuous gradual changes (e.g. dissolution in a phreatic conduit), others occur sporadically and catastrophically (e.g. volcanic explosive activity) and others have a mixed behaviour, with long low-intensity periods punctuated by discrete extreme events (e.g. the flow in a river). The resulting landforms have highly variable persistence times. Some geomorphic features are ephemeral (e.g. small sand dunes), whereas others have a high preservation potential (e.g. glacial cirques).

Landforms can be created by the accumulation of sediments (depositional landforms) or sculpted on pre-existing material (erosional landforms). The genetic interpretation of some landforms may be complicated by morphologic convergence or equifinality, whereby different processes produce very similar features (e.g. fault scarps generated by large gravitational mass movements and tectonic faults). Relict or inherited landforms are those that formed in the past under environmental conditions different than the present-day ones (e.g. an ice-free glacial trough). These geomorphic features can be used to infer paleoenvironmental information and reconstruct the evolution of the landscape. The relative chronology of landform assemblages can be resolved on the basis of spatial relationships (e.g. cross-cutting, inset or superposition relationships) and other attributes, like the degree of degradation and weathering or the maturity of the soils developed on them. These chronological schemes can be significantly improved in some cases utilising geochronological methods.

The structure of the book follows classical subdivisions of geomorphology. The broad field of structural geomorphology, which deals with landforms generated by endogenous processes (active tectonics, volcanism) or controlled by lithological and structural factors, is addressed in five initial chapters. Structural landforms (Chap. 2) are mainly erosional features controlled by the geological structure and the distribution of rocks with different degrees of resistance. Tectonic landforms (Chap. 3) are those created by surface deformation related to active tectonic structures. Volcanic landforms (Chap. 4) include those generated by active volcanism and hydrothermal activity, as well as specific erosional features developed on volcanic rocks. Karst landforms (Chap. 5) occur on soluble rocks and are mainly created by dissolution and precipitation. Granite landforms (Chap. 6) are the geomorphic features that are typically formed on granitic rocks.

A second group of four chapters illustrate landforms associated with widespread geomorphic systems found in most morphoclimatic regions and geological contexts (azonal geomorphology). Weathering landforms (Chap. 7) are related to the breakdown and decomposition of materials at or near the surface by physical, chemical and biological processes. Slope movements (Chap. 8) involve the gravitational displacement of slope-forming materials. Fluvial landforms (Chap. 9) are erosional and depositional features generated by running water, mainly rivers. Coastal landforms (Chap. 10) occur where the land and the sea interact, and waves, tides and nearshore currents play a major role.

The last part of the book comprises three chapters that cover the landforms typically found in specific morphoclimatic regions (climatic geomorphology). Glacial landforms (Chap. 11) are formed by glaciers in cold regions. Periglacial landforms (Chap. 12) are associated with cold non-glacial regions characterised by intense freezing and thawing activity and perennially or seasonally frozen ground. Desert and eolian landforms (Chap. 13) occur in warm arid regions with limited precipitation and scarce vegetation cover.

Structural Landforms

2

Structural landforms are geomorphic features controlled by the underlying geological structure and the distribution of rocks with different resistances to erosion. Erosion tends to preferentially exploit zones underlain by weaker rocks (differential erosion), generating an uneven topography controlled by litho-structural factors. Spatial variations in bedrock erodibility may be related to a number of mappable lithological and structural features. In stratigraphic successions, lithological changes may be related to the superposition of different rock units through conformable or unconformable stratigraphic contacts or to lateral facies changes. Faults (normal, reverse, strike-slip and oblique) may control selective erosion in two main ways. The shattered and sheared material associated with fault planes and fault zones are typically weaker. Faults may also juxtapose different rock units with markedly different resistances to erosion. In areas with plutonic rocks, the lithological changes and their geometrical characteristics may be controlled by the intrusive contact of the igneous bodies with the country rock. The contact may have a dome-shaped geometry (batholiths, laccoliths), a planar subhorizontal shape (sills) or a steep attitude (dykes, volcanic necks).

A major factor controlling the development of erosional landforms in regions underlain by sedimentary successions is the dip of the strata and the distribution of different types of folds (e.g. anticline, syncline, monocline). In these regions, structural surfaces tend to develop on the upper boundary of resistant rock units. These are conformable surfaces in which the topography matches the attitude of the bedding (e.g. flat-topped mesas, dip slopes in cuestas and hogbacks). The orientation and flow direction of drainages with respect to the strike and dip of the strata play an important role in the configuration of these landscapes. In some cases there may be a lack of concordance between the topography and the underlying structure, an example being discordant relief, in which elevated areas occur associated with downthrown fault blocks or synclinal structures consisting of more resistant rocks.

The erodibility of rocks depends on multiple factors in addition to lithology. Fractures and faults are the main mechanical weaknesses in the typically hard igneous rocks. In sedimentary rocks, the main mechanical discontinuities include the bedding planes, joints and faults. The resistance to erosion of the different sedimentary formations is also influenced by other factors including the rock unit thickness, the spacing of the bedding (e.g. massive versus thinly bedded), the presence of interstratified softer beds or the degree of cementation. Some rock types like carbonates, evaporites and granitoid rocks display specific landform assemblages, which are the focus of the chapters on karst and granite landforms.

2.1 Mesa and Butte

The main characteristics of the landscape are frequently controlled by the underlying rock units, which may have contrasting resistance to erosion, and their structure (e.g. horizontal, homoclinal, folded, faulted). Fluvial incision into a flat-lying sedimentary succession with alternating soft and resistant units tends to produce a stepped topography with compound slopes. Structural benches and rock scarps (free-face slopes) develop on resistant rocks, whereas the slopes underlain by the more erodible units are gentler and commonly covered by colluvium (debris slopes) (Fig. 2.1A). Dissection in combination with slope recession may result in the development of residuals or outliers called mesas and buttes. Both are flat-topped hills with steep slopes in which an upper resistant caprock protects the underlying softer sediments. The Spanish term mesa, meaning table, is used for relatively large residuals, whereas butte refers to hills with limited areal extent (Figs. 2.1B, 2.1C and 2.1D).

F. Gutiérrez, M. Gutiérrez, *Landforms of the Earth*, DOI 10.1007/978-3-319-26947-4_2

Fig. 2.1A The Colorado River valley at Dead Horse Point, Utah, shows a stepped topography with structural benches and scarps on flat-lying resistant and nonresistant sedimentary rocks (Published with permission of © Francisco Gutiérrez 2015. All Rights Reserved)

Fig. 2.1B Flat-topped mesa developed on horizontally lying mudstones capped by a resistant gypsum bed (El Planerón, central sector of the Ebro Cenozoic Basin, NE Spain) (Published with permission of © Francisco Gutiérrez 2015. All Rights Reserved)

Fig. 2.1C Butte in Monument Valley, Arizona, consisting of soft sediments (Organ Rock Shale) overlain by resistant scarp-forming rocks, mainly the massive De Chelly Sandstone (Published with permission of © Francisco Gutiérrez 2015. All Rights Reserved)

Fig. 2.1D Butte formed by densely gullied Mancos Shale capped by a relatively thin sandstone bed (west of Hanksville, Utah) (Published with permission of © Francisco Gutiérrez 2015. All Rights Reserved)

2.2 Cuesta and Hogback

Homoclinal structures are sedimentary successions with a constant dip. Where gently dipping sequences consist of stratigraphic units with contrasting resistance, differential erosion of the softer sediments tends to produce a series of asymmetric valleys and ridges parallel to the structural trend. The ridges show a gentle backslope roughly concordant with the resistant strata (dip slope) called a cuesta, a Spanish word that means slope (Fig. 2.2A). The shorter and steeper opposite slope is carved in soft sediments protected by the resistant caprock. Where the strata have a higher dip, the more symmetric ridges controlled by a resistant unit are called hogbacks (Fig. 2.2B). Different terms are used to describe drainages according to

Fig. 2.2A Cuesta topography developed by differential erosion on alternating and gently dipping Jurassic limestone and marl units (contact zone between the Iberian Chain and the Ebro Cenozoic Basin, NE Spain) (Published with permission of © Francisco Gutiérrez 2015. All Rights Reserved)

Fig. 2.2B Hogback on the southern flank of the Gavbast anticlinal ridge, developed on marls of the Gachsaran Formation capped by Guri Limestone (Bastak, Zagros Mountains, Iran) (Published with permission of © Francisco Gutiérrez 2015. All Rights Reserved)

their direction with respect to the dipping structure. Consequent, obsequent and subsequent streams indicate drainages flowing in the dip direction, in a direction opposite to the dip and parallel to the strike, respectively. These terms tend to be replaced by the more intuitive and less confusing words dip, antidip and strike drainages, respectively.

2.3 Chevron and Flatiron

Incision of dipping and well-bedded stratigraphic sequences by evenly spaced transverse gullies produces a serrated pattern given by the layers with different resistance to erosion and colour, called chevrons. These are V-shaped features with the apexes pointing upslope in the interfluves and downstream in the valleys, in accordance with the "Rule of Vs" (Fig. 2.3A). Differential erosion in the interfluves may individualise resistant beds, forming steeply inclined fin-like facets, known as flatirons (Fig. 2.3B).

2.4 Ridge and Rock Wall

Differential erosion in terrains underlain by vertical stratigraphic sequences including rocks with contrasting erodibility commonly results in the formation of ridges along the more resistant beds or formations. These landforms may show a complete morphological gradation, from ridges with a triangular cross profile to tabular walls with a high height-to-width ratio standing up as residuals (Figs. 2.4A and 2.4B). Equivalent landforms develop on vertical dykes where the fracture-controlled igneous intrusive rocks are more resistant than the country rock (Fig. 2.4C). Dykes may form dense swarms with prominent geomorphic expression. They may display parallel, intersecting and radial patterns. The latter generally occur around larger intrusions where the updoming associated with the emplacement of the igneous body produces systems of radial extensional fractures. A less common situation is that of dykes less resistant that the host rock, which favours the

Fig. 2.3A Chevrons developed on well-bedded Late Cretaceous limestones and marls, Bea village, Iberian Chain, NE Spain (Published with permission of © Arturo Polo-Ena 2015. All Rights Reserved)

Fig. 2.3B Flatirons on steeply dipping red conglomeratic sandstone of the Fountain Formation, Boulder, Colorado. These structural landforms occur on the erosional escarpment developed along the eastern margin of the Colorado Front Range (Published with permission of © Francisco Gutiérrez 2015. All Rights Reserved)

Fig. 2.4A Rock walls developed by differential erosion in a formation with vertical attitude consisting of resistant and tabular limestone layers interbedded within softer argillaceous sediments (Finestres, Canelles Reservoir, Spanish Pyrenees) (Published with permission of © Francisco Gutiérrez 2015. All Rights Reserved)

Fig. 2.4B Vertical fin-like ridge developed by selective erosion in sandstone beds of the Fountain Formation, Garden of the Gods, Colorado (Published with permission of © Francisco Gutiérrez 2015. All Rights Reserved)

development of erosional troughs and drainages that exploit the softer intrusives.

2.5 Monoclinal Scarp

A monocline develops where horizontal or very gently tilted strata bend abruptly along a narrow zone to form a step-like fold. These are asymmetric folds with an upper anticlinal hinge, a middle steep limb and a lower synclinal hinge. They are frequently related to ductile bending of strata above steep blind basement faults (fault-propagation folds) and commonly occur in the marginal sectors of structural uplifts and basins (e.g. Colorado Plateau region). They may also form by passive bending over migrating dissolution fronts in dipping evaporitic formations. These folds control the development of monoclinal scarps whose face corresponds to an abrupt dip slope associated with the forward tilted middle limb. This topography may be related to active monoclinal folding itshelf and/or differential erosion (Fig. 2.5). The crest of monoclinal folds is frequently broken by extensional structures like fissures or crestal grabens.

2.6 Anticlinal Ridge

In terrains underlain by folded sedimentary successions, a concordant topography consists of a series of ridges and valleys associated with anticlines and synclines, respectively. The summit of the anticlinal ridges may roughly

Fig. 2.4C Ridge-like erosional remnants controlled by subvertical dykes and protruding from an extensive sand-covered pediplain in the Arabian Shield, Hail area, Saudi Arabia (Published with permission of © Francisco Gutiérrez 2015. All Rights Reserved)

Fig. 2.5 Erosional monoclinal scarp on Mesozoic age formations in the margin of the Uncompahgre Uplift, Colorado National Monument, Colorado (Published with permission of © Jon White 2015. All Rights Reserved)

Fig. 2.6A Anticlinal ridge (Balut-Boland anticline) in Sarvak Limestone, Izeh Zone, Simply Folded Belt, Zagros Mountains, Iran (Published with permission of © Shahram Sherkati 2015. All Rights Reserved)

coincide with the crest of the structure and the slopes with the fold limbs (Fig. 2.6A). These ridges commonly terminate in periclinal structures at the plunging edge or nose of the anticline. Asymmetric anticlines, usually associated with an underlying blind reverse fault or thrust, tend to produce asymmetric ridges with steeper slopes in the forelimb than in the backlimb. In anticlines with a core of softer sediments, once the resistant rocks are eroded from the crest, differential erosion may result in the development of a valley along the axial zone. These erosional depressions bounded by steep slopes are named combe (French terminology). This is a discordant topography in which the depression coincides with the hinge zone of the anticline (valley-centred anticline) (Fig. 2.6B). Drainage may also breach perpendicularly to an anticlinal ridge through a water gap known as cluse (French terminology) (Fig. 2.6C). These transverse drainages may be related to superimposition and/or antecedence.

Fig. 2.6B The Turbón limestone massif (Spanish Pyrenees), an anticlinal structure with an axial valley (combe) excavated into the softer marls and evaporites of its core. Note the periclinal structure on the *left side* of the anticlinal ridge (Image from © Bing Maps)

Fig. 2.6C The Millà anticline, Spanish Pyrenees, breached perpendicularly by the Noguera Ribagorzana River, which has been flooded by the Canelles Reservoir. The resulting water gap is called cluse (Image from © Bing Maps)

2.7 Perched Syncline

In a concordant structural landscape, the valleys coincide with the synclinal structures. However, erosion in the synclinal zone may be slowed down by the presence of a resistant rock unit removed from the adjacent anticlines, where the correlative strata used to occupy higher positions. Progressive entrenchment under these conditions may lead to the development of a perched syncline (discordant relief). The morphology of the hanging syncline depends largely on the inter-limb angle. Tight synclines tend to produce narrow ridges, whereas perched open synclines are commonly expressed as mesa-like landforms with a concave top. The surface drainage and the groundwater flow tend to converge towards the axis of the syncline and the plunging edge of the relief (Figs. 2.7A and 2.7B).

2.8 Conglomerate Pinnacle

These monolith-like structural landforms are conglomeratic pinnacles with vertical cliffs developed by differential weathering and erosion controlled by vertical fractures (Fig. 2.8A). The monoliths may display a wide variety of morphologies (towers, pinnacles, cones, fins) and may reach some hundred metres high (Figs. 2.8B and 2.8C). They occur as isolated residuals or form mazes of monoliths and narrow corridors controlled by the fracture pattern. The formation of conglomerate monoliths requires the concurrence of a number of litho-structural and geomorphic factors, including a thick succession of massive conglomerates; high degree of induration, commonly related to carbonate cementation; vertical fractures with adequate orientation and spacing; and long-term base-level drop typically related to fluvial entrench-

Fig. 2.7A Peña Oroel, a perched syncline with tightly folded conglomerates in the upper part of the relief (Published with permission of © Francisco Gutiérrez 2015. All Rights Reserved)

Fig. 2.7B Perched syncline capped by gently folded Jurassic limestone. Note the stream incised along the axial zone and flowing concordantly with the plunge of the fold. Moncayo Mountain in the background (Iberian Chain, NE Spain) (Published with permission of © Francisco Gutiérrez 2015. All Rights Reserved)

Fig. 2.8A Los Mallos de Riglos, NE Spain, are fracture-controlled conglomerate monoliths developed in calcareous Tertiary alluvial fan facies at the margin of the Ebro Cenozoic Basin, next to the contact with the Pyrenees, Spain (Published with permission of © Francisco Gutiérrez 2015. All Rights Reserved)

Fig. 2.8B The Pisón in Riglos, NE Spain, is a conglomerate monolith with vertical walls more than 300 m high. The pinnacle attached to the cliff is called El Puro (the cigar), a classical challenge for rock climbers (Published with permission of © Francisco Gutiérrez 2015. All Rights Reserved)

ment. Fractures with some deviation from the vertical favour the development of upward tapering pinnacles and cones (Fig. 2.8C). Elongated fins tend to occur in areas with penetrative parallel fractures without any significant conjugate system. Dissolution may play a significant role in the development of these landforms when the conglomerates mainly consist of limestone clasts and a matrix with carbonate cement. Well-developed cave systems have been documented in some conglomerate monoliths.

2.9 Alcove

Alcoves are curved cavities developed in rock cliffs. They are generally related to more rapid recession in soft sediments than in the overlying rocks, which form an overhanging wall or ledge. This differential erosion process is commonly favoured by seepages or springs, whose location tends to be controlled by topographic embayments and the contact between units with contrasting permeability. The issuing water promotes vegetation growth and contributes to weaken the sediments by physical, chemical and biological weathering. Spring and seepage sapping involves cliff undermining (alcove formation), the development of mass movements (alcove degradation) and rapid headward extension of gully heads. Alcoves are very important landforms from the geoarchaeological perspective, since they have been used by humans as shelters and dwellings since prehistoric times (Figs. 2.9A and 2.9B).

2.10 Salt Dome and Namakier

Buried salt, because of its plastic rheology and lower specific weight than other sedimentary formations, tends to flow upward and intrude the overburden rocks (halokinesis and

Fig. 2.8C A cone-shaped conglomerate monolith in the Olgas or Kata Tjuta, central Australia (Published with permission of © Francisco Gutiérrez 2015. All Rights Reserved)

Fig. 2.9A The Long House alcove in Mesa Verde National Park, Colorado. This cavity developed by differential erosion in the gently dipping Cliff House Sandstone was used by the Anasazi for the construction of dwellings, abruptly and mysteriously abandoned in the thirteenth century (Published with permission of © Francisco Gutiérrez 2015. All Rights Reserved)

Fig. 2.9B The alcoves of Santa Orosia, Spanish Pyrenees, are developed in fine-grained beds interbedded within massive conglomerates. The cavities, with seepages, were used for the construction of the La Cueva and San Cornelio hermitages (Published with permission of © Francisco Gutiérrez 2015. All Rights Reserved)

diapirism). Salt flowage is driven by buoyancy and differential loading related to variable overburden thickness and deviatoric tectonic stresses. The development of a deeply entrenched erosional topography may contribute to salt flowage towards the unloaded valleys. The rising salt formations may extrude at the surface forming dome-shaped extrusions where the rate of salt exceeds surface lowering, mainly related to dissolution (Fig. 2.10A). Most salt domes occur in arid or semiarid areas like the Zagros Mountains in Iran. Salt domes may reach more than 20 km in length and 1000 m in local relief, but they may be just the uppermost part of diapiric structures with a feeding vent rooted at more than 10 km depth. Surface dissolution in salt domes produces a cap of insoluble residues that protects the underling salt form further karstification. Differential incision and dissolution, locally controlled by a discontinuous veneer of residual soil and debris, produce striking landscapes of salt pinnacles and pillars (Fig. 2.10C). When the salt domes reach a critical height, the weight-related stress exceeds the yield strength of the salt, so that their lower slopes spread laterally to form namakiers (namak is a Farsi word meaning salt). These are glacier-like salt lobes that override the adjoining country rock as a viscous flow (Fig. 2.10B). The budget and dynamics of salt domes and namakiers are mainly controlled by the rate at which the "viscous fountain" spells salt, salt removal by dissolution and water supplied by rainfall. Namakiers are wasted and retreat due to lower or exhausted salt extrusion and/or more rapid degradation by dissolution. They may generate depositional and erosional features like those produced by glaciers, including moraines, disconnected bodies of "dead salt", cover collapse sinkholes like kettle holes or trimlines.

Fig. 2.10A Khourab Diapir, a dome-shaped salt extrusion largely covered by residual soils in the Zagros Mountains, Iran (Published with permission of © Francisco Gutiérrez 2015. All Rights Reserved)

Fig. 2.10B Dashti Diapir intruded in the core of an anticline, Zagros Mountains, Iran. Gravitational spreading by viscous deformation of the laterally unconfined southern flank of the dome has resulted in the development of a salt flow (namakier) over the southern limb of the anticline (Published with permission of © Francisco Gutiérrez 2015. All Rights Reserved)

Fig. 2.10C Pillar or hoodoo developed on folded salt protected by a boulder of insoluble rock in Dashti Diapir, Zagros Mountains, Iran (Published with permission of © Francisco Gutiérrez 2015. All Rights Reserved)

3 Tectonic Landforms

Tectonic landforms are geomorphic features generated by surface deformation processes related to active tectonic structures. These morphogenetic processes are mainly related to regional stress fields driven by the relative motion of the lithospheric plates; they are particularly intense along the different types of plate margin regions (convergent, divergent and transform or conservative). Crustal deformation associated with plate tectonics is responsible for the main large-scale morpho-structural elements of the Earth, including active orogens, subsiding sedimentary basins, extensive plateaus, oceanic basins, mid-ocean ridges and oceanic trenches. The regions associated with the plate boundaries display the most dramatic tectonic landscapes. In these, vertical and horizontal crustal displacements create the topographic relief that overwhelms the erosion and sedimentation processes. This chapter illustrates the mesoscale tectonic landforms that can be identified and characterised in the field.

Tectonic landforms may be created by faulting (brittle deformation), folding (ductile deformation) or the combination of both, in extensional, compressional or strike-slip tectonic environments. The displacement may be essentially vertical (e.g. dip-slip faults) and horizontal (e.g. strike-slip faults) or have both components. The morphogenetic surface deformation process may be progressive (e.g. creeping faults) or episodic and instantaneous when associated with the sudden displacement on faults during earthquakes (e.g. coseismic surface ruptures). These ground deformation processes may interact with the geomorphic and depositional environments, generating a great diversity of landforms and geomorphic features and anomalies. Tectonic landforms are essential elements used in the identification and mapping of active tectonic structures. They help to constrain the timing of the deformation and give information on strain rates. They also indicate the best sites for conducting paleoseismological studies (trenching), aimed at unravelling the paleoearthquake history of faults and assessing their seismogenic potential. Geomorphology plays an instrumental role in neotectonic and paleoseismological studies. Quaternary landforms are the main markers used to prove and assess the recent activity of tectonic structures, including seismogenic faults.

Gravitational processes related to non-tectonic phenomena are also important. These include large deep-seated slope movements, interstratal evaporite dissolution, salt flow (halokinesis and diapirism) or aquifer consolidation. They may generate landforms such as fault scarps that mimic those developed by active tectonics (morphologic convergence or equifinality). The differentiation between tectonic and non-tectonic deformation processes has relevant implications for seismic hazard assessment. Non-tectonic gravitational faults may cause surface deformation with potential engineering impacts, but they are not capable of producing large damaging earthquakes. However, the correct genetic diagnosis of active structures is not always straightforward, and some of them may have a combined tectonic and gravitational component. It may be difficult to demonstrate whether an active deformation structure penetrates the crust deeply enough to be considered as tectonic and potentially seismogenic. Frequently, the keys for disentangling this dilemma are given by the broad geological and geomorphic context.

3.1 Fault Scarp

Fault scarps are related to surface ruptures caused by faults with some component of vertical displacement (normal, reverse, thrust and oblique faults). Steep dip-slip faults tend to produce straight scarps, whereas scarps related to low-angle faults like thrusts are more sinuous due to low-angle intersection with the ground surface. Active faulting may be expressed at the surface through a single scarp or multiple subparallel scarps usually related to a master fault and several secondary faults. Secondary faults with the same sense of dip as the master fault are called synthetic, whereas those with opposite dip are termed antithetic. Fault scarps commonly have a bow-shaped longitudinal profile, with the highest displacement and height in the central zone, progressively

F. Gutiérrez, M. Gutiérrez, *Landforms of the Earth*, DOI 10.1007/978-3-319-26947-4_3

decreasing towards the fault tips. Arrays of faults may have an en echelon arrangement with overlapping zones called step-overs. These transfer zones between adjacent normal faults may be expressed as accommodation ramps until they rupture and the propagating fault strands join together. Some active faults are characterised by progressive displacement (creeping faults), whereas others have an episodic or stick-slip displacement regime. The latter have the potential to generate damaging earthquakes if their length and rupture area are large enough (seismogenic faults). Sudden normal faulting of unconsolidated deposits commonly creates scarps with a gradient above the repose angle of the deposits. These scarps rapidly degrade by mass movements and erosion to build up a wedge-shaped colluvial deposit below the free face. Colluvial wedges are one of the main evidences of earthquake-related surface faulting used in paleoseismology. As the scarp degrades, the debris slope enlarges at the expense of the retreating free face. Grabens and horsts are the typical structures associated with normal fault systems. Thrust fault scarps may show a higher structural and geomorphic diversity than those related to normal faults. Surface ruptures related to thrust faulting may produce overhanging scarps that readily collapse during seismic shaking, simple pressure ridges underlain by anticlinal structures and complex pressure ridges including crestal normal faults and backthrusts. Repeated faulting over long periods produces fault-controlled range fronts (or mountain fronts) and fault escarpments that may reach hundreds of kilometres in length and a local relief in excess of 1 km. The term fault-line scarp is used for scarps generated by differential erosion where a generally inactive fault juxtaposes rocks with contrasting resistance. These scarps develop on the block with less erodible material, which may be either the footwall or the hanging wall (Figs. 3.1A, 3.1B, 3.1C and 3.1D).

3.2 Fault Facet

The trace of active normal and reverse faults commonly defines two contrasting geomorphic domains. The uplifted block is dominated by erosion and dissection, whereas the downthrown block is frequently a low-relief piedmont area in which the accommodation space generated by tectonic subsidence tends to be filled by aggradation. The boundary between these two morpho-structural domains generally shows a linear break in the slope parallel to the fault called

Fig. 3.1A Undissected normal fault scarps in Dobi Graben, Ethiopia. The height of these young scarps decreases towards the fault tips (Published with permission of © Francisco Gutiérrez 2015. All Rights Reserved)

Fig. 3.1B Fault scarp 4 m high created by the 1915 Pleasant Valley, Nevada, normal fault earthquake (M_s 7.6). In this sector of the 60 km long surface rupture, the fault offsets alluvial–colluvial deposits at the foot of a range front with faceted spurs. The weathered bedrock fault scarp situated above the historical rupture may correspond to a previous coseismic scarp (Published with permission of © James P. McCalpin. All Rights Reserved)

Fig. 3.1C Fault escarpment several hundred metres high associated with the master normal fault that controls the western margin of the Ethiopian Rift, Afar Window, Ethiopia. Note the general inclination of the topography away from the rift in the backtilted footwall (rift shoulder) (Published with permission of © Francisco Gutiérrez 2015. All Rights Reserved)

Fig. 3.1D Range front with a recent fault scarp along its foot related to the active Ragged Mountain thrust, Yakutat microplate, Alaska. The trench in the steep scarp was excavated for a paleoseismological investigation (Published with permission of © Francisco Gutiérrez 2015. All Rights Reserved)

mountain–piedmont junction. The range front or mountain front is commonly dissected by transverse drainages separated by divides truncated by the fault. These facet-shaped truncated spurs are known as triangular or trapezoidal fault facets, depending on their geometry (Figs. 3.2A and 3.2B). They are essentially steep slopes controlled by active faults that are episodically or progressively rejuvenated and reshaped by faulting and erosion. Fault facets are time-transgressive landforms, older in the apex than in the base. Some facets may show geomorphic features indicative of different stages of fault activity and erosion. They may display a compound longitudinal profile, with a steeper basal segment related to recent tectonic rejuvenation. Fault facets affected by dissection may show steep and young facets nested within higher, larger and more degraded facets. Some morphometric indexes derived from mountain fronts and the associated geomorphic systems may be used to roughly assess the slip rate of the range-bounding faults. For instance, rapidly slipping faults tend to produce sharper topography with high and steep facets, straight mountain fronts and narrow and steep valleys in the upthrown block. The lithology and structure of the bedrock also play a major control of the degradation degree of the facets.

3.3 Tectonic Fissure

Tectonic fissures are tensile ground ruptures in which the opposing sides move perpendicular to the failure plane. Tectonic fissures mainly occur on active extensional environments, frequently associated with normal faults, but they can also form in strike-slip and compressional settings. Displacement on dipping normal faults has a vertical (throw) and a horizontal component (heave). The horizontal separation is proportional to the net slip and has a reverse relationship with the fault dip. Horizontal extension may be

Fig. 3.2A Triangular facets on the fault-controlled front of the Sangre de Cristo Mountains, which forms the eastern margin of the San Luis graben, northern termination of the Rio Grande Rift, Southern Rocky Mountains of Colorado (Published with permission of © Vincent Matthews 2015. All Rights Reserved)

Fig. 3.2B Degraded composite facets with small basal truncated spurs nested at the base of older and larger facets, the Concud normal fault, Iberian Chain, NE Spain (Published with permission of © Francisco Gutiérrez 2015. All Rights Reserved)

accommodated by slip along the fault plane and the pulling apart (dilation) of the opposing blocks. Fissures can form in relation to normal faults that refract and steepen near the surface (e.g. listric faults). Horizontal separation in the upper steeper portion of the fault is largely accommodated by the opening of a gap, which may be obliterated by the collapse of the overhanging downthrown block. Extensional fissures may also form along the crest (upper hinge) of monoclinal drape folds related to an underlying normal fault (fault-propagation fold). Tensile fissures may also form in volcanic environments due to local extension above an intruding dyke or by the combined effect of tectonic extension and dyking (volcanic rifts) (Figs. 3.3A and 3.3B). Fissures and fault scarps as much as some kilometres long may also form in relation to human-induced groundwater level declines and the consequent compaction of thick aquifers consisting of unconsolidated sediments. Fissures developed on erodible unconsolidated sediments tend to act as sinks for water and sediments and may rapidly enlarge by focused water erosion (Fig. 3.3C).

3.4 Defeated and Beheaded Drainage

The streams that flow at high angle across tectonic structures are named transverse or discordant drainages. The longitudinal profile of the channels may be disrupted by vertical displacement on active dip-slip faults. In case faulting creates a downstream-facing scarp (downthrown block on the downstream section), the tectonic disruption is expressed as a knickpoint in the channel that tends to be obliterated by fluvial erosion or may propagate upstream by headward retreat. When surface faulting generates an upstream-facing scarp (downthrown block on the upstream side), fluvial erosion tends to remove the weir-like obstacle. The stream will retain its crossflow if bedrock uplift is counterbalanced by fluvial incision (antecedence). Conversely, the upstream-facing scarp may block the water flow generating a defeated stream with a ponded or closed depression on the downthrown block. Downstream of the fault scarp, the channel becomes a beheaded stream, because it is disconnected from its former upstream contributing area, and a hanging wind gap, because

Fig. 3.3A Normal fault scarp and fissure probably associated with the crest of a monocline, Mete Hare Lake, Ethiopian Rift. Fantale stratovolcano in the background (Published with permission of © Francisco Gutiérrez 2015. All Rights Reserved)

Fig. 3.3B Normal fault and fissure related to tectonic extension and magma intrusion in Þingvellir National Park, Iceland. The site is located in the North Atlantic rift system where the North America and Eurasian plates move apart. Skjaldbreidur volcano in the background (Published with permission of © Francisco Gutiérrez 2015. All Rights Reserved)

Fig. 3.3C Fault scarp and earth fissure in Tabah explosion crater (maar), Saudi Arabia. This surface rupture is related to differential compaction of the thick sedimentary fill of the crater due to a dramatic human-induced water level decline caused by groundwater withdrawal. The damage caused by surface deformation accompanied by tremors (small earthquakes) led to the relocation of Tabah village (Published with permission of © Francisco Gutiérrez 2015. All Rights Reserved)

it is elevated with respect to the upstream reach and barely receives any water flow. A similar situation may be found in drainages that flow across growing anticlines (Figs. 3.4A and 3.4B).

3.5 Deflected Drainage

Drainages that are unable to overcome the barrier created by an active tectonic structure like a dip-slip fault or a growing anticline may deflect their course. This type of diversions commonly occurs in terrains with relatively flat topography, since deeply incised drainages can hardly modify their route. A stream may be diverted along the upstream flank of a growing anticline and around its plunging nose. The channel may migrate as the anticline increases its breadth and propagates laterally. Former drainages that used to traverse the anticlinal ridge may be recorded as wind gaps that breach the crest of the ridge. Drainages may also deflect their path in relation to upstream-facing fault scarps. The channel may flow along the foot of the scarp until its termination (fault tip) or a low-relief water gap (Figs. 3.5A and 3.5B).

3.6 Landforms of Strike-Slip Faults

Strike-slip faults are nearly vertical faults in which the blocks move horizontally in opposite directions and approximately parallel to the fault. These faults may laterally offset any geomorphic feature, which may be used as marker to identify and assess the deformation (e.g. drainages, ridges, moraines, terrace scarps). Drainages tend to adapt their path along the fault zone, where the rocks have been weakened by shearing, forming troughs and linear valleys (Fig. 3.6A). Drainages that traverse active strike-slip faults perpendicularly or at high angle commonly display a bayonet-like course with a sharp bend at the fault zone (Fig. 3.6B). This pattern may be related to two mechanisms or the combination of both. An alternative is the lateral offset of the drainage as the blocks move in opposing directions (offset drainage). In this case the horizontal deviation of the drainage should be consistent with the sense of fault motion. Another option is the deflection of the drainage, which adjusts its route to the weakened fault zone along a short stretch (deflected drainage). In this latter case, the fault plays a passive role, providing a band of softer rocks that can be easily exploited by the channel. When strike-slip faults offset a dissected terrain with transverse drainages and intervening ridges, the interfluves on the downstream side may eventually block the drainages to form shutter ridges (Fig. 3.6C). The disruption of the streams may lead to the formation of defeated, offset, deflected and beheaded drainages. Compressional or extensional stresses may be created at bends, step-overs and zones where the fault is split into several anastomosing strands. Uplifts (pressure ridges) and basins (pull-apart basins) tend to form in these restraining and releasing zones dominated by contractional structures and normal faulting, respectively (Fig. 3.6D).

Fig. 3.4A Normal fault escarpment with a pond related to a defeated stream in the downthrown block and a perched beheaded valley (wind gap) on the footwall (eastern sector of the Ethiopian Rift between Awash and Dire Dawa) (Published with permission of © Francisco Gutiérrez 2015. All Rights Reserved)

Fig. 3.4B Transverse drainages disrupted by an upstream-facing normal fault scarp. The scarp has blocked the drainage creating defeated streams and flood-prone closed depressions on the upstream side and perched beheaded streams on the downstream side. Note the prominent triangular facets in the interfluves (Zenzano Fault, Iberian Chain, NE Spain) (Published with permission of © Francisco Gutiérrez 2015. All Rights Reserved)

Fig. 3.5A Drainage that traverses an active graben bounded by upstream- and downstream-facing scarps. The channel is deflected along the sediment-filled graben and escapes from the downthrown block through a water gap associated with a cross fault, Zagros Mountains, Iran (Published with permission of © Francisco Gutiérrez 2015. All Rights Reserved)

Fig. 3.5B River deflected around the plunging nose of an active anticline north of Deh Dasht, Zagros Mountains, Iran (Image from © Google Earth)

Fig. 3.6A Trough with ponded depressions attributable to local extension along the trace of the right-lateral Boconó fault in the Mucubají Lake area, Venezuela (Published with permission of © Franck Audemard 2015. All Rights Reserved)

Fig. 3.6B Laterally offset transverse drainages on the right-lateral San Andreas Fault, Carrizo National Monument, California. The bayonet-like sharp bend of the channels indicates that the block on the upper right side of the fault moves towards the reader (Image from © Bing Maps)

Fig. 3.6C Shutter ridge and offset-deflected drainage associated with a left-lateral oblique fault. The block on the downslope side moves up and to the left. The upslope-facing scarp of the shutter ridge has partially blocked a drainage, which is offset and deflected to the right. Note elongated ponded troughs at the foot of the fault scarp. The paleoseismological trench excavated across the scarp revealed a fault plane with oblique slickensides (Martin Lake, Cordova area, Yakutat microplate, Alaska) (Published with permission of © Francisco Gutiérrez 2015. All Rights Reserved)

Fig. 3.6D Pressure ridge (foreground) in a local zone of compression associated with the right-lateral Boconó fault in the Mucubají Lake area, Venezuela. The trace of the fault runs along the foot of the ridge. Faulted Pleistocene moraine in the background (Published with permission of © Franck Audemard 2015. All Rights Reserved)

3.7 Flexural-Slip Fault Scarp

Flexural slip refers to a folding mechanism whereby the flexing of sedimentary rocks is accompanied by bedding-parallel slip on mechanically weak contacts between layers. This mechanism is illustrated by the slippage of the pages in a telephone book when folded. These are peculiar faults with no stratigraphic separation whose net displacement is proportional to the thickness of the sedimentary packages and the angle of the fold limb. Moreover, the slip decreases to zero towards the hinge line. Flexural slip is common in stratigraphic sequences including stiff and ductile layers and bedding planes with low shear strength. Flexural slip in areas affected by active folding may be expressed as tracts riddled with scarps, troughs and benches parallel to the strike of the underlying beds. For instance, in slopes underlain by steeply dipping beds, outward toppling of the strata produces uphill-facing scarps (antislope scarps) and small half-graben depressions of their upslope side, which may disrupt drainages and act as sediment traps (Figs. 3.7A and 3.7B). In contrast, inward rotation of the strata related to flexural slip may be expressed as a series of downhill-facing scarps and benches. Flexural-slip faults are difficult to recognise due to the lack of stratigraphic separation and because the associated geomorphic anomalies may be also attributed to differential erosion. They are best displayed when the folded strata are overlain unconformably by a younger unit that ruptures by the subjacent flexural slip.

Fig. 3.7A General view of slopes riddled with flexural-slip antislope scarps and troughs related to active folding, south of Martin Lake, Cordova area, Yakutat microplate, Alaska (Published with permission of © Francisco Gutiérrez 2015. All Rights Reserved)

Fig. 3.7B Steep slope with two fresh-looking uphill-facing flexural-slip scarps with enclosed linear depressions on their upslope side, Martin Lake area, Yakutat microplate, Alaska (Published with permission of © Francisco Gutiérrez 2015. All Rights Reserved)

4 Volcanic Landforms

Volcanic eruptions involve the surface emission of magma (molten igneous material with gases) in subaerial or submerged environments; the accumulated volcanic products are mainly lava and pyroclasts. From the geomorphological perspective, these are constructional processes that lead to the creation of oceanic crust and mid-ocean ridges, seamounts, volcanic islands, large stratovolcanoes or newly emerged land. The topography built by the accumulation of volcanic rocks and deposits tends to be tear down by erosional and collapse processes, including the development of calderas and giant landslides formed when major portions of volcanic edifices fail (flank collapses). The magma may emerge through localised conduits (central eruption), along fissures or aligned vents (fissure eruption). There are two basic eruptive modes: the emission of lava (effusive) and the ejection of pyroclastic material (explosive). The latter may include molten or solidified magma and rock fragments from previous volcanic deposits or the basement. Low-viscosity magmas with low gas contents (mainly mafic/basic magmas) tend to produce effusive eruptions, whereas gas-rich viscous magmas (mainly felsic/acidic magmas) are conductive with explosive activity. Viscous magmas may block the vent or the crater, and the consequent overpressure may be released through violent explosions. The ejected pyroclastic material may be transported and deposited in various ways. Fine-grained particles (ash) may rise vertically within the hot air of the eruptive columns reaching as much as tens of kilometres high. The ash may be transported long distances by air currents and accumulate as ash fall deposits that drape the ground surface. Large particles including blocks (solid) and bombs (partially molten) tend to follow ballistic trajectories. Explosive eruptions may also produce glowing pyroclastic flows. They may be related to violent explosions induced by trapped pressurised gas or the collapse of the eruptive column due to insufficient buoyancy. Explosive eruptions can also be driven by the interaction of magma with groundwater, which rapidly transforms into pressurised gas (phreatomagmatic eruptions). The magma may also interact with surface water bodies (hydromagmatic eruptions), including deep ocean water, lakes, ice caps or glaciers on the upper parts of stratovolcanoes, controlling the eruptive style and leading to the development of secondary hazardous processes including lahars or jökulhlaups.

The distribution of young volcanoes, the composition of their magmas and the eruptive styles are largely controlled by plate tectonics. Most of the volcanoes, including those built in the submarine realm, occur associated with convergent and divergent plate margins. In the convergent or destructive plate boundaries, where there is subduction of oceanic lithosphere, magmas are typically viscous, have a high content of volatiles and produce explosive eruptions (e.g. Antilles, Andes). These subduction-zone volcanoes account for more than 80 % of the documented historical eruptions, including the most disastrous ones. Volcanic eruptions associated with divergent or constructive plate margins (mid-ocean ridges) are commonly dominated by an effusive behaviour producing by far the highest volume of erupted magma and the largest number of active submarine volcanoes. There are also a significant number of volcanoes in intraplate settings; these include the volcanic islands related to hot spots (e.g. Hawaii, Canaries) or volcanoes associated with continental rifts (e.g. the East African Rift Valley).

Each year, around 60 of the approximately 550 active volcanoes on Earth experience eruptions. These produce a wide variety of hazardous processes, including pyroclastic flows, lava flows, pyroclastic falls, large slope movements, lahars, explosions, gas emissions and tsunamis induced by explosions or large landslides. The detrimental effects of the eruptions can be widespread. They can affect air traffic services with flight suspensions due to ash clouds. They also cause disastrous indirect consequences including famine related to damaged crops covered by ash fall deposits. About 260,000 people have died directly as a result of volcanic eruptions since 1600 AD. Since 1900, pyroclastic flows and lahars have accounted for a great proportion of the fatalities related to volcanic disasters. In the nineteenth century, two major volcanic eruptions in Indonesia caused more than 120,000 fatalities (1815 Tambora, 1883 Krakatoa). In 1902, a pyro-

F. Gutiérrez, M. Gutiérrez, *Landforms of the Earth*, DOI 10.1007/978-3-319-26947-4_4

clastic flow from Mt. Pelée, Martinique, devastated Saint-Pierre killing around 29,000 people. In 1985, lahars (volcanic mud flows) generated by the eruption of the Nevado del Ruiz, Colombia, caused some 25,000 fatalities, mainly in Armero. In Saint-Pierre, it is said that because of an election the mayor of the city did not allow the people to evacuate in the days prior to the eruption. In Armero, evacuation was not activated, despite the advanced prediction of a mud flow and the town receiving warnings about the approaching lahar. A considerable number of large cities are located within areas that may be affected by hazardous volcanic processes (Naples, Quito, Tokyo, Managua, Yogyakarta). Fortunately, our ability to predict the behaviour of volcanic eruptions using information from the geological record and monitoring systems has increased substantially in recent decades. Despite the hazards, volcanic activity has highly beneficial effects, including geothermal energy, formation of ore deposits, fertile volcanic soils, raw materials and highly attractive landscapes. Some volcanoes and volcanic landscapes are among the most visited touristic attractions worldwide (e.g. Mount Fuji, Yellowstone National Park).

4.1 Volcano

Volcanoes show a wide range of sizes and shapes conditioned by a number of factors, including the composition of the magma and its gas content, the eruptive styles, the characteristics of the plumbing system, the amount and rate of magma supply, the presence of water beneath or on the surface and the constructional and erosional history of the volcano. Long-sustained eruptive activity on a stable vent may produce large volcanic edifices named stratovolcanoes. They commonly consist of numerous layers of lavas and pyroclast with considerable syndepositional dip piled up during multiple eruptions (polygenetic volcanoes). Stratovolcanoes may reach more than 3 km in local relief, and in some volcanic islands, they are just the tip of towering piles of volcanic rocks rising as much as 10 km above the ocean floor (Hawaii, Canaries). These volcanoes are often crown by craters hundreds of metres across that may be occupied by water or lava lakes (Figs. 4.1A, 4.1B and 4.1C). A lateral shift in the active vent, which may be related to the excessive height of the volcano or the plugging of a previous conduit by solidified

Fig. 4.1A The Teide Volcano (3718 m a.s.l.), Tenerife, Canary Islands, is a compound stratovolcano that rises around 7500 m above the adjacent seafloor. The last eruption of this volcano occurred in the Early Middle Ages. The Teide Volcano is included in the UNESCO World Heritage List (Published with permission of © Francisco Gutiérrez 2015. All Rights Reserved)

Fig. 4.1B One of the craters of the Irazú Volcano, Costa Rica, partially filled with a lake. The last eruption in this stratovolcano, with an elevation of 3431 m a.s.l., occurred in 1963 (Published with permission of © Francisco Gutiérrez 2015. All Rights Reserved)

Fig. 4.1C Cotopaxi stratovolcano, Ecuador. This volcano that reaches 5911 m a.s.l. has small glaciers on its summit and has erupted around 50 times since 1738 (Published with permission of © Tom Rayner 2015. All Rights Reserved)

Fig. 4.1D Tuya or table mountain related to a subglacial eruption, south of Myvatn Lake, northern Iceland (Published with permission of © Francisco Gutiérrez 2015. All Rights Reserved)

magma, may lead to the formation of a compound volcano with parasitic cones. Shield volcanoes are edifices with gentle convex slopes and a broad and relatively flat summit. Their morphology reflects the prevalence of low-viscosity lavas that are able to flow long distances over gently sloping terrain. Subglacial eruptions, common in Iceland, generate peculiar volcanoes influenced by the presence of an ice cap and meltwater. They produce flat-topped and steep-sided volcanoes (tuyas or table mountains) related to the accumulation of volcanic products in a lake generated by subglacial melting (Fig. 4.1D).

4.2 Cone

These are relatively small volcanoes generally formed by a single eruptive episode (monogenic). They typically display a conical or horseshoe shape, may have a crater in the top and reach hundreds of metres in height and more than 2 km in basal diameter. There are two main types of volcanic cones, in order of abundance: scoria cones and tuff cones. Scoria cones, also known as cinder and tephra cones, form by pyroclastic eruptions of low-viscosity, generally basaltic magmas (Figs. 4.2A and 4.2B). The cones mainly consist of scoriaceous lapilli (2–64 mm), bombs (>64 mm) and minor ash, mainly deposited ballistically and by lava fountains (Strombolian/Hawaiian eruptions). The coarse-grained deposits commonly show crude bedding and limited amount of lithic clasts, and the highly vesicular particles may be welded (spatter). Tuff cones generate by explosive phreatomagmatic eruptions related to the interaction between the ascending magma and groundwater or surface water. In contrast to scoria cones, they mainly consist of well-bedded, fine-grained tuff (harden ash) with a large proportion of lithic clasts (country rock fragments).

Fig. 4.2A Cone-shaped (Caldera del Corazoncillo) and horseshoe-shaped (Montaña Rodeos) scoria cones in the Timanfaya National Park, Lanzarote, Canary Islands, Spain. The cone on the right has a basal diameter of approximately 800 m and is around 60 m high. These cones surrounded by basaltic lava formed during the 1730–1736 fissure eruption, which covered one third of the island, leading to its temporary abandonment (Image from © Google Earth)

Fig. 4.2B Eldfell scoria cone in Heimaey, Westman Islands, Iceland. The 200 m high cone formed in the 1973 eruption, during which lava flows (on the left of the image), destroyed part of the village and considerably enlarged the size of the island. The flows were about to close the harbour of this village, whose economy mainly relies on income from the fishing industry (Published with permission of © Francisco Gutiérrez 2015. All Rights Reserved)

4.3 Maar

Maar is a German word used to designate craters excavated in the country rock by violent phreatomagmatic eruptions. These depressions are typically surrounded by an annular rampart of ejected material with a steeper inner slope. The craters are generally subcircular and may reach 3 km in diameter and 500 m in depth. The rim deposits are typically less than 50 m thick and dip away from the crater. They are mostly composed of well-bedded, finely brecciated country rock particles (lithics). Maar craters commonly host lakes that behave as long-lasting sediment sinks. They may contain continuous stratigraphic records covering wide time spans of great interest from the paleoenvironmental perspective (Figs. 4.3A, 4.3B and 4.3C).

Fig. 4.3A Graenavatn maar in Krisuvik, southwestern Iceland. The crater is 350 m across and 46 m deep (Published with permission of © Francisco Gutiérrez 2015. All Rights Reserved)

Fig. 4.3B Tabah maar crater surrounded by a tuff ring, Saudi Arabia. The thick sedimentary fill the depression generated by the phreatomagmatic explosion constitutes a prolific aquifer used for the cultivation of palm trees (Published with permission of © Francisco Gutiérrez 2015. All Rights Reserved)

Fig. 4.3C Aci Gol maar in Karapinar area, central Anatolia, Turkey (Published with permission of © Francisco Gutiérrez 2015. All Rights Reserved)

4.4 Lava Dome

Lava domes are mounds that form by the slow extrusion of viscous magma from a volcanic vent. The outer part of the extrusion rapidly cools and solidifies providing sufficient strength to hold high and steep slopes. They may display a wide variety of shapes from steep-sided to tabular (spine-, dome-, mushroom-, platelike), mainly depending on the extrusion rate, the yield strength of the lava and its cooling rate. The growth of lava domes may last from days to several decades and may reach several kilometres in diameter and hundreds of metres high. They frequently form in the crater of volcanoes, but may also develop from vents controlled by fractures, like the rim of a caldera. The surface is frequently blocky, with many cracks related to brittle deformation of the rigid crust, and may display explosion pits. Lava flows related to the lateral spreading of domes (coulées) show a corrugated surface with arcuate pressure ridges whose convexity points in the flow direction (ogives). The oversteepened flanks of growing domes may eventually collapse resulting in the development of catastrophic avalanches of glowing lava fragments (pyroclastic flows). These pyroclastic flows may travel long distances and reach high speeds, especially when they originate in the summit of large stratovolcanoes (Figs. 4.4A, 4.4B and 4.4C).

Fig. 4.4A Montaña Rajada, a Holocene phonolitic lava dome 450 m across developed at the foot of the Teide Volcano, in the landslide-related Cañadas depression, Tenerife, Canary Islands (Published with permission of © Francisco Gutiérrez 2015. All Rights Reserved)

Fig. 4.4B Little Glass Mountain, a ca. 1000-year-old obsidian-rich rhyolite dome (background) with blocky lava flows, Klamath National Forest, California (Published with permission of © Donald L. Crain 2015. All Rights Reserved)

Fig. 4.4C Rhyolite domes related to fracture-controlled eruptions younger than 5000 years old, Mono Craters, Mono–Inyo volcanic chain, Inyo National Forest, California (Published with permission of © Francisco Gutiérrez 2015. All Rights Reserved)

4.5 Lava Flow and Associated Features

The lava emitted from a volcanic vent or a fissure tends to flow downgradient as a viscous fluid until it solidifies. The morphological characteristics of lava flows depend on a number of factors such as the viscosity and gas content of the magma, the topography, the emission rate and the total amount of lava. Low-silica mafic lavas (e.g. basalt) have much lower viscosity than the high-silica felsic lavas (e.g. rhyolite). Low-viscosity lavas may erupt within an enclosed crater and form a temporary lava lake (Fig. 4.5A). These basic lavas are highly mobile and may run over long distances even in low-gradient areas. Two main types are differentiated according to the surface texture, which are designated by Hawaiian words. Pahoehoe flows display relatively smooth surfaces with features related to the plastic deformation of the cooling upper part of the lava, including lobes, ropelike corrugations and bulbous features (Figs. 4.5B and 4.5C). In contrast, aa flows are characterised by a jagged surface of angular clinker-like decimetre-scale blocks, related to the brecciation of the upper brittle crust of the flow (Fig. 4.5D). Aa lava flows typically display well-defined lateral levees and may produce accretionary lava balls. These are blocks of lava that roll over the flow surface enlarging their size by a snowball effect and may travel some distance ahead of the lava flow front (Fig. 4.5D). The internal structure of mafic lavas may display well-defined columnar jointing, consisting of regular prisms with five to seven sides. The change from a molten to a solid state of the lava involves a volume reduction which is accommodated by contraction through regular joints that tend to be perpendicular to the cooling edges. Lava flows that solidify in channels or lava tubes may develop peculiar radial jointing to form a rosette (Figs. 4.5E, 4.5F and 4.5G). The surface of low-viscosity basic lava flows may display hornitos (Spanish word for little oven) and tumuli. Hornitos are protruding chimneys or pinnacles commonly with an open pipe up to a few metres high. They are related to the upward spurt of lava through cracks open in the upper crust of the flow when it overwhelms the conduit. Tumuli are blister-like domes related to the upwarping of the upper crust of a lava flow related to the pressure exerted by the underling flowing lava (Fig. 4.5I). High-viscosity felsic lavas travel short distances, tend to be thicker and generally display a blocky surface with large pressure ridges (ogives) and steep frontal and lateral edges (Figs. 4.5A, 4.5B, 4.5C, 4.5D, 4.5E, 4.5F, 4.5G, 4.5H and 4.5I).

4.6 Lava Tube

Rapid cooling and solidification of the upper surface and edges of lava flows may produce a tube-like rock crust, while hot and fluid lava keeps on flowing along the pipe. The crust has an insulating effect that retards the cooling of the flowing lava, which may completely evacuate the conduit to form a lava tube (Fig. 4.6A). Lava tubes are essentially the same age as the rocks and are commonly associated with recent pahoehoe basaltic flows with gentle gradient. They typically lie a few metres below the surface and their linear low-sinuosity pattern is controlled by the path of the lava flows, which in turn is governed by the topography. Lava tubes may reach more than 60 km in length, and the collapse of their roof produces numerous pseudokarst sinkholes known as jameos in the Canaries and pukas in Hawaii (Fig. 4.6B). Lava tubes commonly have arched cross sections and irregular floors with fragments of broken lava. Some tubes are partially eroded into previous lava flows by

Fig. 4.5A Solidified lava lake in the crater of the Buyoyo scoria cone in northern Fuerteventura, Canary Island (Published with permission of © Francisco Gutiérrez 2015. All Rights Reserved)

Fig. 4.5B Pahoehoe lava in the Kilauea Volcano, Hawaii (Published with permission of © Juan Carlos Carracedo 2015. All Rights Reserved)

Fig. 4.5C Pahoehoe lava flow with ropy texture in the Malpais National Monument, Tularosa Basin, New Mexico (Published with permission of © Francisco Gutiérrez 2015. All Rights Reserved)

Fig. 4.5D Aa basaltic lava flow dating back to the Middle Age with a rough blocky surface on a steep slope of the Teide Volcano, Tenerife, Canary Islands. Note dark accretionary lava balls accumulated on light-coloured felsic pyroclasts beyond the front of the flow (Published with permission of © Francisco Gutiérrez 2015. All Rights Reserved)

Fig. 4.5E Columnar jointing in basalts on a coastal cliff, Dyrhólaey, southern Iceland (Published with permission of © Francisco Gutiérrez 2015. All Rights Reserved)

Fig. 4.5F Transversal exposure of columnar joints in a basaltic flow defining prisms with 5–6 sides, Dyrholaey, southern Iceland (Published with permission of © Francisco Gutiérrez 2015. All Rights Reserved)

Fig. 4.5G A basaltic flow with radial jointing (rosette), Tenerife, Canary Island (Published with permission of © Francisco Gutiérrez 2015. All Rights Reserved)

Fig. 4.5H Hornito formed by lava welling up from an underlying tube, flank of Kilauea Volcano, Hawaii Volcanoes National Park (Published with permission of © Arthur Palmer 2015. All Rights Reserved)

Fig. 4.5I Tumuli on the surface of a lava flow next to Mete Hare Lake, at the foot of Fantale stratovolcano, Ethiopia (Published with permission of © Francisco Gutiérrez 2015. All Rights Reserved)

Fig. 4.6A Cueva del Viento lava tube, Icod de los Vinos, Tenerife, Canary Islands. This 18.5 km long cave system formed ca. 27 kyr ago in basaltic lava flows derived from Pico Viejo, a parasitic cone associated with the Teide Volcano (Published with permission of © Alfredo Lainez 2015. All Rights Reserved)

Fig. 4.6B Sinkholes related to the collapse of a lava tube, Brunahraun, southern Iceland (Published with permission of © Francisco Gutiérrez 2015. All Rights Reserved)

melting, rather by abrasion. They may have a wide range of secondary deposits related to solidification of dripping or flowing lava, precipitation of minerals from hot gases, weathering and accumulation of autochthonous and allochthonous detrital material. Molten material dripping from the ceiling of the caves may produce lava stalactites and stalagmites (Figs. 4.6A and 4.6B).

4.7 Caldera

Caldera, a Spanish word that means big cooking pot, refers to large circular or elliptical depressions related to the subsidence of the rocks overlying a magma chamber. The removal of material from the magma reservoir related to significant eruptions causes the subsidence of its roof due to loss of basal support. Large calderas, as much as tens of kilometres across, are typically related to big explosive (ash flow) eruptions. Small calderas a few kilometres in diameter may also form in the crest of basaltic and andesitic volcanoes. Caldera subsidence commonly occurs during and soon after the eruptions and may be accommodated by three main deformation styles, which should be considered as end members: Collapse of piston-like blocks with limited internal deformation and bounded by ring faults. The bounding faults are commonly vertical or have a steep inward dip and may reach kilometre-scale throws. Sagging of the undermined rocks, leading to the development of a basin structure with centripetal dips. Chaotic brecciation and collapse, which typically occurs in small calderas. Calderas, especially those related to collapse subsidence, are commonly bounded by arcuate escarpments. These steep slopes are essentially the scarps related to the master ring faults, largely modified by mass movements. The caldera depressions may be filled with syn- and post-collapse sequences several kilometres thick with a high proportion of pyroclastic deposits. Some calderas may undergo post-collapse uplift related to renewed rise of magma. This resurgence phenomenon may include the doming of the depression floor and the formation of protruding intrusions. The large caldera-forming explosive eruptions, although infrequent, are one of the most catastrophic geological events. The occurrence of such phenomenon in the near future would cause difficult to foresee disastrous societal and environmental consequences in our planet (Figs. 4.7A, 4.7B and 4.7C).

Fig. 4.7A The ca. 600 m deep Crater Lake in Oregon occupies a 7700-year-old caldera around 10 km across with a resurgent andesitic intrusion (Wizard Island) (Image from © Bing Maps)

Fig. 4.7B The Sete Cidades caldera in Sao Miguel Island, Azores. This 5 km wide caldera, related to the collapse of the summit of a stratovolcano, has a peripheral wall around 500 m high and several nested cones (Image from © Google Earth)

Fig. 4.7C The summit caldera of Mauna Loa volcano, Hawaii, formed around 1000–1500 years ago, is 6.2 km long and 2.5 km wide. The peripheral scar reaches 170 m in local relief and the floor is occupied by a solidified basaltic lava lake (Published with permission of © Tim Lawnicki 2015. All Rights Reserved)

4.8 Flank Collapse

The flanks of large volcanic edifices may experience gravitational collapse leading to the development of gigantic landslides (see Chap. 8). These flank collapses include some of the largest documented slope movements on Earth, with volumes of the order of thousands of km^3. The gravity-driven instability may produce a wide range of landslides, ranging from coherent rockslides to rapid and long run-out flow-type movements (e.g. debris avalanches, debris flows, turbidites). Flank collapses are particularly common in volcanic oceanic islands, especially in their early stages of development (e.g. Canaries, Hawaii). The headscarp of the structural collapses is typically expressed in the landscape as conspicuous amphitheatre-like escarpments which may control the shape of volcanic islands. The landslide deposits may reach extremely high run-outs, especially when they accumulate on the seafloor. Debris avalanches and debris flows may form lobes hundreds and thousands of kilometres long, respectively. The former are characterised by a rough and hummocky topography including huge protruding blocks. Large-volume and extremely rapid flank collapses that form on the coast and/or in the seafloor have the potential to create hazardous impulse water waves (tsunamis). The main predisposing factors involved in the development of these large landslides include the growth and oversteepening of the volcanic edifices by successive eruptions and the presence of alternating competent and weak layers with high syndepositional dip. Some of the factors that may play a significant role in the initiation and development of flank collapses include the intrusion of dykes (rift zones) and cryptodomes, the weakening of rocks by hydrothermal alteration, tectonic and volcanic earthquakes and sea-level drop. Intrusions may oversteepen the slopes, induce upward and lateral stresses, create and dilate discontinuity planes and increase pore water pressure (Figs. 4.8A, 4.8B and 4.8C).

4.9 Volcanic Neck

Volcanic necks, also known as volcanic plugs or pitons, are erosional remnants related to the differential erosion of the material that surrounds the highly resistant rock that fills a relict volcanic vent (Fig. 4.9A). The plug may consist of solidified lava or hardened agglomerate. The softer material removed by erosion may correspond to a pre-existing volcanic cone associated with the plug or the country rock in which the vent opened its way. Glacial action may create steep erosional scarps along the flanks and upflow side of the plug, and a streamlined tail of glacial deposits on the downflow side, generating a landform described in Scotland as crag and tail (e.g. Edinburgh Castle) (Figs. 4.9B and 4.9C).

4.10 Volcanic Hydrothermal Manifestations

The heat from igneous bodies in volcanic areas (e.g. magma chamber) is transferred to the surface by conduction and convection. Conduction involves transmission of heat

Fig. 4.8A The peculiar shape of El Hierro Island in the Canaries, Spain, is controlled by three flank collapses. Each embayment corresponds to the headscar of a landslide 50–150 km^3 in volume, whose deposits have been identified in the adjacent seafloor. The major axis of the island is around 29 km (Image from © Google Earth)

Fig. 4.8B Partial view of the 16 km long arcuate escarpment of Las Cañadas (Tenerife, Spain), which corresponds to the headscar of a multi-episodic flank collapse, occurred around 200 kyr ago. The erosional depression generated by this ca. 750 km^3 mega-landslide has been partially filled by the Teide Volcano. Phonolitic lava domes and flows (coulées) in the foreground. Image taken from the eastern flank of the Teide Volcano (Published with permission of © Francisco Gutiérrez 2015. All Rights Reserved)

Fig. 4.8C Curved escarpment corresponding to the headscarp of a large flank collapse in Cofete, Jandía Peninsula, Fuerteventura, Canary Islands. Mantled pediments between the steep cliffs and the coast (Published with permission of © Francisco Gutiérrez 2015. All Rights Reserved)

through rocks without mass transfer, whereas the heat transferred by moving fluids (liquid, gas) through permeable rocks is termed convection. The convective fluids that discharge at the surface give rise to various manifestations with characteristic depositional, geomorphic and alteration features. Such fluids may have a magmatic and/or meteoric origin. The subsurface gaseous phase of water is called vapour that turns into steam at the surface as it partially condensates. Magmatic fluids commonly have aggressive acidic gases (e.g. HCl, SO_2, HF) with a high alteration capability. Solfataras are emissions of sulphurous gases and steam that may deposit large amounts of sulphur around their vents. Fumaroles refer to vents that discharge vapour turning into steam with condensed droplets at the surface. The discharge of hot water in relatively flat areas may produce depressions with hot, often boiling, water or water-saturated mud (hot water pools and hot mud pools) (Figs. 4.10A and 4.10B). The mud is commonly derived from the alteration of rocks by hot and low pH magmatic fluids. Geysers eject violently superheated water and steam, often with some periodicity. The subsurface structure of geysers comprises a plumbing system with a constriction close to the surface. As the water fills the system and rises, steam forms and expands building up the pressure. At a critical point, the confined bubbles lift the water above causing the geyser to initially overflow. The decrease in pressure eventually leads to sudden boiling and the resulting large amounts of steam force water to violently erupt out of the vent. Geysers commonly deposit a mound of silica-rich sinter called geyserite (Figs. 4.10C and 4.10D).

Fig. 4.9A Needle Rock, Crawford, a volcanic neck in the Southern Rocky Mountains of Colorado (Published with permission of © Vincent Matthews 2015. All Rights Reserved)

Fig. 4.9B A steep-sided plug of phonolitic trachyte with a tail of glacial deposits on the downflow side, generated by differential erosion and deposition during the Quaternary ice sheet glaciation, North Berwick Law, Midlothian, Scotland (Published with permission of © Francisco Gutiérrez 2015. All Rights Reserved)

Fig. 4.9C The Midi d'Ossau Peak is a volcanic plug consisting of Late Paleozoic andesitic rocks, French Pyrenees. Note the talus cones that grade into a lobate rock glacier, nested behind a grass-covered moraine (Published with permission of © Francisco Gutiérrez 2015. All Rights Reserved)

Fig. 4.10A Hot water pool connected to a vent covered by geyserite in Yellowstone National Park, Wyoming. Steaming conical mounds of silica-rich sinter (geyserite) in the background (Published with permission of © Francisco Gutiérrez 2015. All Rights Reserved)

Fig. 4.10B Pools of boiling mud in an area strongly affected by hydrothermal alteration, Yellowstone National Park, Wyoming (Published with permission of © Francisco Gutiérrez 2015. All Rights Reserved)

Fig. 4.10C Geysir, southwest Iceland, is the pulsating spring that gives name to this infrequent type of hydrothermal manifestation (Published with permission of © Francisco Gutiérrez 2015. All Rights Reserved)

Fig. 4.10D Geyser erupting from a narrow orifice surrounded by geyserite, Yellowstone National Park, Wyoming (Published with permission of © Francisco Gutiérrez 2015. All Rights Reserved)

Karst Landforms

5

The word karst comes from a limestone region that straddles Slovenia and Italy called *kras* in Slovenian or *carso* in Italian. Karst refers to terrains in which the geomorphological and hydrological features, both at the surface and in the subsurface, are largely governed by the dissolution of carbonate and/or evaporite rocks. These environments are widely distributed, since carbonate and evaporite rocks occur over about 20 % of the Earth's ice-free continental area. Dissolution commonly overrules mechanical erosion and produces distinctive landforms such as internally drained closed depressions, rock outcrops sculpted by a wide variety of karren and caves. The hydrology is characterised by the prevalence of underground drainage through conduits and caves largely related to the dissolutional enlargement of discontinuity planes. Surface water readily infiltrates in landscapes in which the surface drainage network is typically absent or deranged, and it flows rapidly in the subsurface through conduit networks and caves towards springs. Three main types of karst settings can be differentiated: bare karst, covered or mantled karst and interstratal karst, depending, respectively, on whether the soluble rocks are exposed at the surface, covered by unconsolidated deposits, or overlain by non-karst rocks (caprocks), respectively.

There are some important differences between carbonate and evaporite karst systems with relevant implications for landform development. Evaporites are much more soluble and dissolve more quickly than carbonates. The equilibrium solubilities of gypsum ($CaSO_4 \cdot 2H_2O$) and halite, the most common evaporite minerals, are 2.4 gL^{-1} and 360 gL^{-1}, respectively. By comparison, the solubilities of calcite ($CaCO_3$) and dolomite ($MgCa[CO_3]_2$) in normal meteoric waters are generally lower than 0.1 gL^{-1}. This explains why solution features form much more rapidly in evaporites and why flowing water holds its aggressiveness for longer distances. In limestone areas, most of the dissolution occurs in a thin shallow zone called the epikarst, where the water rapidly losses its dissolution capacity. Gypsum and halite have a significantly lower mechanical strength and a more ductile rheology than carbonate rocks. This explains why instability phenomena (e.g. cave collapse) tend to operate at higher rates in evaporite areas and sagging is a common subsidence mechanism. Moreover, evaporite sequences may include hypersoluble rocks (e.g. halite, K-chlorides) rarely exposed at the surface. The interstratal dissolution of these commonly unrecognised units may produce large non-tectonic subsidence structures with geomorphic expression (e.g. graben depressions, monoclinal-fold scarps).

One of the most common morphological characteristic of karst landscapes is the lack of a well-developed surface drainage network and the presence of enclosed depressions. Large depressions as much as several tens of kilometres long, with flat floors and typically controlled by the structural grain, are called poljes. Sinkholes are smaller depressions with internal drainage widely regarded as one of the main landforms diagnostic of karst. They are also designated with the term doline, coming from the Slavic word *dolina*. Sinkholes display a wide range of geometries (cylindrical, conical, bowl or pan shaped). Their diameter and depth are commonly of the order of metres or tens of metres, although there are gigantic sinkholes more than 500 m deep and wide. Compound depressions resulting from the coalescence of adjoining sinkholes are known as uvalas. Two main groups of sinkholes can be differentiated from the genetic perspective: solution sinkholes and subsidence sinkholes. Solution sinkholes are generated by differential corrosional lowering of the ground surface where karst rocks are exposed or mantled by thin soil. Subsidence sinkhole refers to the various types of sinkholes that result from subsurface dissolution and downward gravitational movement (deformation, internal erosion) of the undermined overlying material. These are the most problematic sinkholes from the ground instability perspective. Subsidence sinkholes are classified in this book according to the subsidence mechanism (collapse, suffosion, sagging) and the type of material affected by downward gravitational displacement (cover, bedrock and caprock). Cover refers to unconsolidated deposits, bedrock to soluble karst rocks and caprock to non-karst rocks. Frequently, the development of subsidence sinkholes is related to more than

F. Gutiérrez, M. Gutiérrez, *Landforms of the Earth*, DOI 10.1007/978-3-319-26947-4_5

one subsidence process, which may affect different types of material (complex sinkholes). Other common geomorphic features found in soluble rocks are karren. These are small-scale dissolution forms developed on bare rock surfaces (free karren) or beneath a soil cover (covered karren). Karren is classified by combining morphologic and genetic criteria into circular forms and two groups of linear forms: hydraulically controlled and structurally controlled.

Most known caves have formed by meteoric water circulating in unconfined aquifers (epigenic caves). Distinguishing characteristics separate cave sections developed above the water table (vadose conditions) from those formed within the saturated zone (phreatic conditions), where the conduits are entirely filled by water. According to some authors, hypogenic caves are those formed in karst systems dominated by upward groundwater flow, whereas others use the term for situations in which the chemical aggressiveness of the groundwater is derived from below the surface. Caves display a huge variety of depositional and solutional features. The most common mineral deposits that form in caves are mainly carbonates and are collectively known as speleothems. They can be grouped according to the hydrodynamic condition of the water from which they precipitate and the mechanism responsible for attaining supersaturation conditions: pool, dripping, flowing and capillary water, plus evaporation. The dissolution features developed on bedrock surfaces of caves are called speleogens.

Karst processes and landforms pose multiple environmental and geotechnical problems. Karst waters are extremely vulnerable to pollution, due mainly to the direct connection between the surface and the underlying aquifers and the fast circulation of the groundwater. It is estimated that around a quarter of the global population depends on karst water for their supply. The main hazards and geotechnical problems endemic to karst include sinkholes, uneven rockhead, reservoir leakage and the flooding of underground excavations.

5.1 Polje

The Slavic word polje indicates flat and alluviated fields. The geomorphological term refers to large closed depressions with conspicuously flat floors developed in karst, mainly in carbonate terrains. They may reach more than 50 km in length and as much as 700 km^2 in area. These topographic basins are generally elongated and oriented concordantly with the structural grain. The margins of poljes are commonly formed by steep limestone slopes. In some cases, the bottom is surrounded by staircased corrosion surfaces that record alternating periods of deepening and planation. Poljes typically have internal drainage, mostly achieved by infiltration through swallow holes (ponors). Some poljes are affected by flooding when the inflow of water by run-off and springs exceeds the outflow capacity of the ponors or when the water table rises above the ground surface. These enclosed basins may be captured by the external drainage network and become open poljes. The bottom of poljes are commonly underlain by a thin veneer of unconsolidated deposits, mostly alluvium and clay-rich karstic residues (terra rossa) that mantle the limestone bedrock. The alluviated surface may be interrupted by residual hills called hums. The origin of the extremely flat floors in these poljes is related to corrosion processes controlled by the position of the water table. In some cases, poljes correspond to tectonic basins bounded by active faults. The floor of these morphostructural depressions may be underlain by a thick sedimentary fill. This is an example of morphologic convergence or equifinality, whereby different processes (corrosion planation vs. tectonically induced aggradation) may produce similar landforms (Figs. 5.1A and 5.1B).

5.2 Solution Sinkhole

Solution sinkholes are generated by differential corrosional lowering of the ground surface where karst rocks are exposed or merely soil mantled. These depressions are commonly shallow and their bottom may be filled by unconsolidated deposits including clay-rich karstic residue (terra rossa) and detrital material derived from the erosion of the margins (Fig. 5.2). The development of solution sinkholes is favoured by preferential flow and dissolution in higher permeability zones within the epikarst. The spatial distribution of solution sinkholes tends to be associated with structural discontinuities (e.g. joints and faults), more permeable and soluble stratigraphic units, and along the contacts between karst and non-karst rocks (contact karst). In humid tropical regions, the land surface may be totally riddled with densely packed solution sinkholes and intervening hills, often conical in shape, forming an egg-box-like topography. This peculiar karst landscape in which the divides between adjoining sinkholes form a mesh pattern is known as polygonal karst (e.g. the cockpit of Jamaica).

5.3 Collapse Sinkhole

The formation of collapse sinkholes requires the presence of underground cavities with unstable roofs. A cavity in bedrock or caprock may propagate upwards through successive roof collapses, until it intercepts the land surface producing a bedrock collapse sinkhole or a caprock collapse sinkhole, respectively (Figs. 5.3A and 5.3C). The upward stoping of the cavity may generate large chimneys filled with breakdown deposits called breccia pipes. Cover deposits may

Fig. 5.1A Niequsi Polje, Croatia. Poljes, with extensive arable lands and springs, constitute the only suitable areas for human settlement in some regions of the Dinaric karst (Published with permission of © Francisco Gutiérrez 2015. All Rights Reserved)

Fig. 5.1B Dabarsko Polje in Bosnia–Herzegovina is an elongated depression bounded by steep escarpments controlled by active faults (Published with permission of © Francisco Gutiérrez 2015. All Rights Reserved)

Fig. 5.2 Solution sinkhole with a thin fill of residual and detrital deposits used for cultivation (Villar del Cobo village, Iberian Chain, NE Spain) (Published with permission of © Francisco Gutiérrez 2015. All Rights Reserved)

Fig. 5.3A Bedrock collapse sinkholes (Zenzano, Iberian Chain, NE Spain). The largest depression is 170 m on the long axis (Published with permission of © Francisco Gutiérrez 2015. All Rights Reserved)

Fig. 5.3B Cover collapse sinkhole induced by water pumping from limestone aquifers in a farming area (Al Jouf, Saudi Arabia) (Published with permission of © Francisco Gutiérrez 2015. All Rights Reserved)

migrate downwards into conduits and voids developed in the underlying soluble bedrock. When the cover is cohesive and dominated by brittle rheology, internal erosion of the mantling deposits results in the development of a soil cavity. Upward migration of the void by successive soil failures eventually leads to the formation of a cover collapse sinkhole (Fig. 5.3B). Another less common collapse style is the foundering of the cavity roof as a rigid body with limited internal deformation and controlled by annular subvertical faults. Collapse sinkholes are commonly subcircular depressions with steep or overhanging edges and may occur catastrophically. Their volume provides a minimum estimate for that of the underlying cavity. The largest documented collapse sinkhole in the world is Xiaozhai Tiankeng (Heavenly Pit), with a major length of 626 m.

5.4 Cover Suffosion Sinkhole

Cover deposits may migrate downwards into voids associated with the rockhead of the soluble bedrock (internal erosion). When the undermined cover behaves as soft or loose granular material, it may settle gradually by continuous deformation (without the development of significant failure planes) generating funnel- or bowl-shaped cover suffosion sinkholes, typically a few metres in diameter (Fig. 5.4).

5.5 Sagging Sinkhole

Sagging refers to the downward flexing of ductile sediments related to a lack of basal support by subsurface dissolution. In mantled karst settings, the progressive lowering of the rockhead by differential dissolution may lead to the gradual sagging of the cover producing cover sagging sinkholes (Fig. 5.5). Similarly, interstratal karstification within the soluble bedrock may be accompanied by the concomitant passive bending of the overlying rock strata, resulting in the development of bedrock and/or caprock sagging sinkholes. Sagging produces basin structures with centripetal dips that only affect the sediments situated above the karstification zone (disharmonic structure). This subsidence mechanism is particularly frequent in areas underlain by evaporitic formations including highly soluble salt beds. Sagging sinkholes are typically large, shallow and diffusely edged depressions. Their development does not require the formation of cavities.

Fig. 5.3C Caprock collapse sinkhole formed in basalts underlain by karstified limestone (Alesaweah, Al Jouf area, Saudi Arabia) (Published with permission of © Francisco Gutiérrez 2015. All Rights Reserved)

Fig. 5.4 Active cover suffosion sinkhole developed in the *bottom* of a large bedrock collapse sinkhole. This funnel-shaped depression, resembling the cone that forms in an hourglass, is related to the downward migration of the gravelly cover into cavities within the underlying bedrock (Villar del Cobo, Iberian Chain, NE Spain) (Published with permission of © Francisco Gutiérrez 2015. All Rights Reserved)

Fig. 5.5 Quaternary alluvium (mantled pediment deposits) affected by sagging structures related to the karstification of the underlying salt-bearing evaporitic bedrock (exposed in excavations for the Madrid–Barcelona high-speed railway in the vicinity of Zaragoza city, Ebro Cenozoic Basin, NE Spain) (Published with permission of © Francisco Gutiérrez 2015. All Rights Reserved)

5.6 Evaporite Collapse Valley

The subjacent karstification of evaporites may cause the subsidence of the overlying rocks and the land surface, producing large-scale morphostructures and depressions. A striking example is given by the evaporite collapse valleys of the Canyonlands section of the Colorado Plateau in Utah and Colorado. These depressions have formed along anticlines with salt bodies in their cores as much as 4 km thick (salt walls). Interstratal karstification of the evaporites has caused the subsidence of the anticline crests and the development of depressions along the fold axes. The subsidence depressions may exceed 50 km in length and are bounded by escarpments several hundred metres high (Fig. 5.6A). Two subsidence styles have been recognised. Ductile sagging of the anticline crest, which generates an axial synform flanked by paired anticlines or monoclines. More frequently, the collapse of the anticline crest is controlled by systems of steep normal faults parallel to the folds axis. Active synthetic and antithetic normal faults in the margins of the depressions produce lateral grabens as well as uphill- and downhill-facing scarps that locally disturb drainages (defeated and beheaded streams, knickpoints) (Fig. 5.6B). Some of these salt valleys, not related to fluvial erosion but to dissolution-induced subsidence, are traversed perpendicularly by major superimposed rivers, such as the Colorado River in Spanish Valley (Utah) or the Dolores River in the Paradox Valley (Colorado) (Figs. 5.6A and 5.6B).

5.7 Tower Karst

This is a dramatic karst landscape of residual carbonate hills protruding from a surrounding corrosion plain, mostly found in humid tropical areas. The hills may reach hundreds of metres in height, and their morphology may vary from steep-sided towers to cones with more gentle slopes (often called cone karst). These erosional remnants are called mogotes in some sites of the Caribbean, including Cuba and Puerto Rico. The surrounding plains are cut across carbonate bedrock, typically mantled by a cover of unconsolidated deposits. Chinese geomorphologists distinguish between isolated towers (fenglin) and clusters of hills with a common bedrock base (fengcong) (Figs. 5.7A and 5.7B). Tower karst is considered by a number of geomorphologists as a more evolved stage than polygonal karst. Vertical corrosional lowering of the depressions and the consequent increase in height of the intervening hills dominates when the water table is deep. Once the topographic surface is close to the water table and the epikarst merges with the phreatic zone, the development and expansion of corrosion plains prevails. The residual hills may display caves and horizontal notches at different elevations recording the progressive or episodic lowering of water table and the surrounding plain. The slopes and cliffs of these residuals are frequently indurated by secondary carbonate (case hardened) favouring their preservation and the development of steep slopes (Figs. 5.7A and 5.7B).

5.8 Corrosion Plain

These are flat and nearly horizontal erosional surfaces cut across soluble bedrock. Corrosional plains are frequently veneered by alluvium resting on a planar bedrock strath that truncates the pre-existing geological structures. The development of these surfaces by corrosion planation is controlled by a shallow water table (base-levelled plains). Dissolution by aggressive water from rain and snow melting is mostly concentrated at a shallow depth in the epiphreatic zone, causing the removal of irregularities and the expansion of the plains by the retreat of the adjacent uplands. These plains mainly evolve by backwearing erosional pro-

Fig. 5.6A View of a portion of Paradox Valley developed along the Paradox salt anticline and traversed perpendicularly by the Dolores River. The morphostructural depression is related to interstratal dissolution of the salt core and the collapse of the anticline crest. Note the fresh-looking graben and fault scarps at the margins of the collapse valley with associated disturbed drainages and enclosed depressions. The light-coloured exposures in the valley floor correspond to the gypsum-rich caprock of the Paradox salt formation (Image from © Bing Maps)

Fig. 5.6B Recent graben related to subjacent evaporite dissolution on the NE margin of the Paradox Valley. Densely jointed Jurassic Entrada Sandstone in the foreground (Published with permission of © Francisco Gutiérrez 2015. All Rights Reserved)

Fig. 5.7A Tower karst with isolated towers (fenglin) and groups of residual hills with a common base (fengcong) in Yangshuo, Guilin area, China (Published with permission of © Scott Weatherson 2015. All Rights Reserved)

Fig. 5.7B Groups of conical residual hills (fengcong) in Libo, South China (Published with permission of © Paul Williams 2015. All Rights Reserved. Copyright)

cesses rather than downwearing, similarly to pediplains. Consequently, there are time-transgressive surfaces, progressively younger towards the foot of the retreating slopes. The development of extensive corrosion plains requires long periods of tectonic quiescence and a relatively stable water table (Fig. 5.8).

5.9 Fluviokarstic Canyon

The incision of fluvial systems by dissolution and mechanical erosion in carbonate bedrock tends to produce deep and narrow canyons or gorges. The walls of these impressive bedrock channels may reach more than 1 km in height. The development of these gorges is favoured by the high mechanical resistance of carbonate rocks and the limited surface run-off that restricts mass wasting processes. Fluviokarstic canyons are frequently associated with allogenic drainages (substantial run-off derived from non-karst areas) that flow across carbonate terrains affected by tectonic uplift (antecedence) or significant base-level drop. The fluviokarstic canyons perched above the water table behave as influent or losing streams. These are difficult settings from the construction of dams and reservoirs due to the high potential for water leakage problems. The incisional history of karst canyons may be inferred through the investigation and dating of multilevel cave systems that record the position of former base levels (Fig. 5.9).

Fig. 5.8 Corrosion plain cut across limestone in Kistanje area, close to the Krka River, Croatia (Published with permission of © Francisco Gutiérrez 2015. All Rights Reserved)

5.10 Blind Valley

Blind valley is a valley in which the flow infiltrates underground through one or several stream sinks (also known as swallow holes or ponors). The sinks may be related to the formation of collapse sinkholes in the stream bed. The local base-level drop associated with the formation and enlargement of the sink may induce the entrenchment of the channel upstream through the propagation of an incision wave. Downstream from the swallet there is typically an abandoned valley section beheaded by and perched above the sink. During flood events, when stream discharge exceeds the absorption capacity of the sink, the water may eventually overflow along the downstream valley segment. The lower end of blind valleys commonly displays an arcuate upstream-facing scarp with a ponor at its foot and an incision in the crest corresponding to the abandoned hanging valley (Figs. 5.10A and 5.10B).

5.11 Pocket Valley

Pocket valleys or steepheads are the reverse of blind valleys, since they form in association with large springs at the foot of a cliff. The retreat of the cliff is accelerated by spring sapping and produces an amphitheatre-like steephead or pocket valley. The water that issues at the spring undermines the base of cliff favouring the development of mass movements, mainly rock falls, resulting in rapid headward recession. Pocket valleys may be alternatively related to the progressive collapse of the roof of the underground channel that feeds the spring (Figs. 5.11A and 5.11B).

5.12 Karrenfield

Karren is a German word used to designate small-scale dissolution forms developed on bare rock surfaces (free karren) and beneath a soil cover (covered karren). Alternative terms include lapiaz and lapiés. Karren can be considered as the uppermost part of the epikarst, a relatively thin zone in the shallow subsurface in which the soluble rocks show a high degree of karstification. The processes involved in karren development include biokarstic corrosion, wetting from dew, direct rainfall, sheet and channelled flow as well as infiltration and percolation. Outcrops of karst rocks, including gypsum and salt, may be riddled with assemblages of karren, forming scenic karrenfields. The terms microkarren and karren are commonly used for features with a characteristic

Fig. 5.9 The Mont-rebei Canyon in the Spanish Pyrenees has been carved by the Noguera Ribagorzana River in a thick succession of Late Cretaceous limestone. The river is a transverse drainage that crosses perpendicularly the fold and thrust belt of the southern Pyrenees. Here, the river is a superimposed and antecedent drainage. The notch on the right wall is an artificially excavated path (Published with permission of © Francisco Gutiérrez 2015. All Rights Reserved)

dimension smaller or larger than 1 cm, respectively. Karren can be classified combining morphological and genetic criteria into circular forms and two groups of linear forms, hydraulically controlled and structurally controlled (Fig. 5.12).

5.13 Circular Karren

There are four main types of karren with subcircular plan geometry. Solution pits are small hollows, typically 1–5 cm across, with concave bottoms and circular or elliptical geometry in plan view. They may form densely packed honeycomb-like complexes (Fig. 5.13A). Solution pans (kamenitzas) are closed depressions with flat and subhorizontal floors up to a few metres in diameter. They commonly have scarped edges, which may display basal corrosion notches and overflow channels (Fig. 5.13B). Heelprints (trittkarren) occur on sloping rock surfaces. They are planar corrosion bevels opened downslope and enclosed by a steep cirque-like backscarp. The backscarps are commonly a few centimetres in height and may be fluted. The bevels are generally more than 10 cm wide (Fig. 5.13C). Solution shafts or pipes are tube-like hollows with circular or elliptical section that form beneath a soil cover. They may display multiple orientations and form interconnected complexes with branching and tortuous trajectories (Fig. 5.13D).

Fig. 5.10A Zalmska blind valley and Biogradski ponor in Nevesinje Polje, Bosnia–Herzegovina (Published with permission of © Francisco Gutiérrez 2015. All Rights Reserved)

Fig. 5.10B Ephemeral stream in the flat-topped summit of Mount Sedom salt diapir (Israel) interrupted by a stream sink. The swallow hole corresponds to a collapse sinkhole connected to a cave system (Published with permission of © Francisco Gutiérrez 2015. All Rights Reserved)

Fig. 5.11A Pocket valley associated with Cetina Spring, Croatia (Published with permission of © Francisco Gutiérrez 2015. All Rights Reserved)

Fig. 5.11B The steephead associated with the Urederra Spring forms a conspicuous embayment in the Urbasa limestone massif, western Spanish Pyrenees. Image from © Bing Maps

Fig. 5.12 Karrenfield in Pag Island, Croatia. The limestone is densely sculptured by multiple karren, including grikes, rillenkarren and shafts (Published with permission of © Francisco Gutiérrez 2015. All Rights Reserved)

Fig. 5.13A Solution pits on limestone, Cap de Formentor, Pollença, Mallorca Island, Spain (Published with permission of © Ángel Ginés 2015. All Rights Reserved)

Fig. 5.13B Solution pans or kamenitzas in Kistanje, Croatia (Published with permission of © Francisco Gutiérrez 2015. All Rights Reserved)

Fig. 5.13C Assemblage of heelprints or trittkarren developed on a steep limestone slope in Zalmska River, close to the Biogradski ponor, Bosnia–Herzegovina (Published with permission of © Francisco Gutiérrez 2015. All Rights Reserved)

Fig. 5.13D Solution shafts or pipes developed on Neogene lacustrine limestone close to La Tranquera Reservoir, Iberian Chain, Spain (Published with permission of © Francisco Gutiérrez 2015. All Rights Reserved)

5.14 Linear Karren (Hydraulically Controlled)

The most common types of dissolution channels that develop on bare rock surfaces are solution flutes and solution runnels. Solution flutes or rillenkarren are probably the most common karren type that occur on sloping rock exposures. They are tightly packed straight and parallel channels that head at the crest of rock slopes and extinguish downslope. The flutes, typically a few millimetres to some centimetres wide, display parabolic cross profiles and are separated by sharp ridges. They are generated by direct rainfall (Figs. 5.14A and 5.14B). Solution runnels or rinnenkarren are steep-sided and round-bottomed channels larger than solution flutes and separated by wide interfluves. They form on rock slopes where unconfined sheetflow is partitioned and concentrated into Hortonian channels. The width and depth of the runnels tend to increase downslope and sinuosity shows an inverse relationship with the topographic gradient. Rundkarren is similar to rinnenkarren but forms beneath a soil cover and has more rounded shapes (Figs. 5.14A, 5.14B, and 5.14C).

5.15 Structural Karren

Structural karren collectively designates linear karren resulting from the dissolutional enlargement of discontinuity planes (e.g. joints, faults, bedding planes). Grikes or kluftkarren typically develops along subvertical planes, either exposed or soil covered. They taper downwards due to the rapid reduction in the aggressiveness of the percolating waters. The grikes filled with soil or regolith are named cutters (Fig. 5.15A). Intersecting grikes related to conjugate systems of fractures isolate blocks termed clints (flachkarren). Recession of clint tops related to the widening of the surrounding grikes ultimately leads to the development of pinnacles (spitzkarren). Exposed and subsoil pinnacles are typically fluted by different types of solution channels (Figs. 5.15B and 5.15C). Karrenfields dominated by grikes on subhorizontal surfaces are called pavements, which are best developed in areas with a planar surface carved by glaciers or wave action (Fig. 5.15D). Long-sustained widening of fractures by corrosion, in combination with other weathering and erosion processes, may result in the development of large corridors or labyrinths (giant grikes) with intervening towers (Figs. 5.15E and 5.15F).

Fig. 5.14A Rillenkarren developed on salt with interbedded expansive clays (Cardona salt diapir, Ebro Cenozoic Basin, NE Spain) (Published with permission of © Francisco Gutiérrez 2015. All Rights Reserved)

5.16 Phreatic Cave

Most known caves have formed by meteoric water circulating in unconfined aquifers. These epigenic caves are typically a three-dimensional complex of conduits with a tree-like or branchwork pattern. Conduits converge as tributaries towards progressively lower and larger passages feeding a major underground trunk channel connected to a spring. The morphological characteristics of caves largely depend on whether they form under vadose or phreatic conditions, that is, above or below the water table, respectively. Under phreatic conditions, the conduits are entirely filled by flowing water and dissolution acts around the entire perimeter. Consequently, phreatic caves tend to have a tube-like geometry with circular or ellipsoidal cross section (Fig. 5.16). Tubes controlled by discontinuity planes may have ellipsoidal sections related to more rapid widening along the plane. Phreatic passages mostly form at or below the water table and commonly show a subhorizontal distribution controlled by the elevation of the spring or base level. The episodic entrenchment of the base level (e.g. fluvial incision) may result in the development of multilevel caves, with progressively older fossil cave levels at higher elevations. These levels in some cases can be correlated with fluvial terraces to allow the reconstruction of the long-term incision history of landscapes (Fig. 5.16).

5.17 Vadose Cave

Vadose caves form by water flowing downwards above the water table, commonly guided by fractures and stratification planes. These caves are characterised by a sig-

Fig. 5.14B Rillenkarren sculptured on a steep limestone slope (Santa Elena, Gállego Valley, Spanish Pyrenees) (Published with permission of © Francisco Gutiérrez 2015. All Rights Reserved)

nificant vertical component including shafts and canyon-like passages. Shafts are well-like vertical cavities formed by water flowing downwards along fractures (Fig. 5.17A). In inclined vadose passages, unlike phreatic conduits, dissolution and erosion are focused on the floor, generating canyon-like underground channels. Phreatic caves may change into vadose conditions by a water table drawdown, commonly related to fluvial incision and base-level drop. The floor of the tubes may be incised by vadose flows generating passages with a keyhole-like cross section, comprising an upper subcircular or ellipsoidal phreatic portion and a subsequent basal vadose channel (Fig. 5.17B). Once the phreatic caves become air filled, some mineral deposits such as dripstones or flowstones may start to form.

5.18 Hypogenic Cave

The adjective hypogenic is commonly used for caves formed in karst systems dominated by upward flow, including artesian aquifers. Some authors restrict the term to situations where the chemical aggressiveness of the water is produced below the surface. Such increase in the karstification potential could be related to multiple factors, including water mixing, cooling of thermal waters or the incorporation of deep-sourced gases like CO_2 or H_2S. The latter gas transforms into sulphuric acid where it reaches oxidising conditions. Hypogenic caves formed by rising water commonly lack any relationship with the surface recharge or topography and display characteristic patterns like network or spongework mazes, isolated irregular chambers or giant phreatic

Fig. 5.14C Rinnenkarren developed on a steep limestone slope carved by glacial erosion (Linza, Ansó Valley, Spanish Pyrenees). Note the progressive downslope widening and deepening of the solution channels (Published with permission of © Francisco Gutiérrez 2015. All Rights Reserved)

Fig. 5.15A Soil-filled grikes (cutters) controlled by discontinuity planes and intervening clints. This is an artificial exposure of the epikarst. Limestone quarry near Zavala, SW of Popovo Polje, Bosnia–Herzegovina (Published with permission of © Francisco Gutiérrez 2015. All Rights Reserved)

Fig. 5.15B Fluted pinnacles in the Stone Forest (Shilin), Kunming, Yunnan, China. These are exhumed pinnacles sharpened by surface dissolution (Published with permission of © Scott Weatherson 2015. All Rights Reserved)

shafts. The main mesoscale morphological indicators of hypogenic speleogenesis include basal feeder conduits, ceiling channels and clusters of ceiling pockets such as cupolas and domes (Fig. 5.18). Vadose features like shafts and canyons are commonly absent. In confined settings, the cave features are related to slow phreatic flow. In unconfined conditions, condensation corrosion above the water table, accelerated by the incorporation of acidic gases, may play a significant role in shaping the morphology of the caves.

5.19 Flank-Margin Cave

Along the seacoast, the freshwater in the phreatic zone forms a wedge that floats on the denser saline water. The landward dipping freshwater–saltwater interface is called the halocline. The dissolution capacity of the freshwater is enhanced or renewed at the halocline by mixing with the saltwater, mainly due to the saline or foreign-ion effect. The increase in aggressiveness is particularly important along the seaward edge of the halocline, where mixing is more effective. The resulting mixing corrosion leads to the development of flank-margin caves, which tend to form close and parallel to the coastline and show limited relationship with the surface karst features. These caves are typically small, with spongework patterns, and frequently display rounded ceilings with cupolas. Flank-margin caves are common in islands with recent marine limestones (e.g. the Caribbean), and they may form perched subhorizontal clusters raised by tectonic uplift (Fig. 5.19).

5.20 Pool Speleothems

The mineral deposits that form in caves are collectively known as speleothems. They can be grouped according to the hydrodynamic condition of the water from which they precipitate and the mechanism responsible for attaining supersaturation conditions: pool, dripping, flowing and capillary water, plus evaporation. Precipitation of calcium carbonate in pools is commonly related to the loss of CO_2 to the cave air, which is most active at the water surface. Crystals may precipitate on the submerged surface forming crusts, typically with a laminated internal structure. On irregular

Fig. 5.15C The Pinnacles of Nambung National Park, Cervantes, Western Australia, are not related to structurally controlled dissolution but to preferential cementation of aeolian sands along discrete vertical zones, probably guided by roots, and subsequent differential erosion (Published with permission of © Francisco Gutiérrez 2015. All Rights Reserved)

Fig. 5.15D Pavement at Malham Cove, Yorkshire, UK (Published with permission of © Anthony H. Cooper 2015. All Rights Reserved)

Fig. 5.15E Corridors hundreds of metres long mainly controlled by a penetrative ENE-WSW trending fracture system on Cretaceous limestones and dolomites (Ciudad Encantada, Iberian Chain, NE Spain) (Image from © Google Earth)

Fig. 5.15F Ruiniform landscape with limestone towers isolated by well-developed grikes and corridors (Torcal de Antequera, Betic Cordillera, Southern Spain) (Published with permission of © Francisco Gutiérrez 2015. All Rights Reserved)

Fig. 5.16 Djatlo Cave, Bosnia–Herzegovina, with an upper ellipsoidal section generated under phreatic conditions and guided by subhorizontal bedding (Published with permission of © Francisco Gutiérrez 2015. All Rights Reserved)

bedrock surfaces, crusts grow more rapidly on inward projections, leading to the formation of coalescing botryoidal knobs called mammillary crusts or cave clouds (Fig. 5.20A) Rimstones are dam-like deposits that form on the edge of pools, commonly designated as gours. Rimstones typically have a downstream convex geometry and a steeper upstream face. They grow more rapidly in the vertical direction due to higher turbulence and degassing along the crest of the dam (Figs. 5.20B and 5.20D). Shelfstones are flat-topped ledges of calcium carbonate that grow inwards along the surface of calm and stable pools. They can also form around other partially submerged speleothems like dripstones (Figs. 5.20C, and 5.20D).

5.21 Dripstones

These are speleothems related to dripping water that commonly have internal layering. Stalactites are elongated deposits with an internal conduit that grow downwards from the cave ceiling or overhanging surfaces (Fig. 5.21A). The water flows through the inner tube and deposits carbonate in the stalactite tip. Water may also flow and accumulate minerals along the outer sides of the stalactite causing its widening. These stalactites typically taper downwards. Straw stalactites form when precipitation is restricted to the lower opening. These are small diameter and straight stalactites with an effective inner drain enclosed by a thin and sometimes translucent wall (Fig. 5.21B). Stalagmites are pillars or mounds that grow upwards from the cave floors where dripping water falls persistently (Fig. 5.21C). The water drops splash and may be projected away or flow over the sides of the stalagmites (Fig. 5.21D). Columns form when a stalactite merges with the underlying stalagmite. Draperies are curtain-like deposits that form on overhanging walls by water trickles that descend attached to the lower edge of the drapery by surface tension. Many draperies merge in the downflow edge with stalactites. They may be sinuous, thin and translucent (Fig. 5.21E). Cave pearls are a peculiar speleothem type whose formation is also related to dripping water. These are oncoids that form in shallow pools agitated by dripping water. Detrital particles are progressively coated by concentric layers of calcium carbonate. The oscillations of the coated grains induced by dripping water favour the development of spheroidal particles and prevent their amalgamation. They may also attain roller shapes when precipitation occurs on elongated particles (Fig. 5.21F).

Fig. 5.17A Vadose shaft in Mount Sedom salt diapir, Israel (Published with permission of © Francisco Gutiérrez 2015. All Rights Reserved)

5.22 Flowstones

Flowstones are generated by precipitation from water running as a thin sheetflow on sloping surfaces. They display a wide variety of morphologies (sheets, mounds, cones) that spread across walls and floors (Fig. 5.22). The irregularities in the flowstone surfaces tend to be accentuated with time due to higher turbulence and degassing and more rapid precipitation (shelf-accelerating process). The upward growth of irregularities oriented perpendicular to the flow direction may result in the development of rimstones and gours. Flowstones may become perched ledges or canopies if the sediments on which they grew are eroded away.

5.23 Speleothems Formed by Capillary Water

These speleothems are related to small amounts of water that move slowly governed by capillary and surface tension forces through minute pores and fissures or as thin films. Water movement and mineral deposition are not controlled by gravity, and consequently the speleothems may grow in multiple and capricious directions. Helictites are elongated precipitates that may show tortuous and branching geometries (Fig. 5.23A). They grow where water seeps from tiny voids or fissures in bedrock or other speleothems. They have a minute inner conduit through which the water moves by

Fig. 5.17B Passage with an upper ellipsoidal section and a lower vadose incision, Travers du Four Cave, Le Clapier, Aveyron, France (Published with permission of © Jean-Yves Bigot 2015. All Rights Reserved)

Fig. 5.18 Amalgamated ceiling cupolas generated by condensation corrosion related to convection above the water table, Des Serpents Cave, France (Published with permission of © Philippe Audra 2015. All Rights Reserved)

Fig. 5.19 Image of Cueva del Alemán, Mona Island, Puerto Rico, showing the characteristic cuspate ceiling of flank-margin caves (Published with permission of © John Mylroie 2015. All Rights Reserved)

Fig. 5.20A Partially eroded mammillary crust with concentric layering in Crnulja Cave, connected to one of the main ponors in Popovo Polje, Bosnia–Herzegovina (Published with permission of © Francisco Gutiérrez 2015. All Rights Reserved)

Fig. 5.20B Rimstones and gours in Cueva del Caserío de Seso, Spanish Pyrenees (Published with permission of © Juan Carlos Gordillo 2015. All Rights Reserved)

capillary forces and probably also hydrostatic pressure. The small size of the tube precludes the formation of discrete droplets in the helictite tip and water is permanently attached to the surface by adhesive forces. Shields consist of two parallel subcircular or ellipsoidal plates separated by a narrow gap. Their radial growth is related to slow water seepage from the attachment point towards the edge of the crack in the perimeter of the shield (Fig. 5.23B).

5.24 Speleothems Formed by Evaporation

Evaporation in caves, commonly favoured by air currents, leads to mineral precipitation and induces capillary water seepage towards drier areas. Different minerals may precipitate sequentially as water evaporates, and the ionic concentration increases and supersaturation conditions are reached for minerals with different solubilities. Coralloid is a general term that embraces a wide range of nodular and fingerlike deposits. Cave popcorn refers to small isolated or coalescing balls that grow attached to bedrock surfaces or other speleothems (Fig. 5.24A). Cave corals are irregular fingerlike speleothems that grow upward, commonly in windy sectors of caves (Fig. 5.24B).

5.25 Collapse Breccia and Debris Cone

Cave ceilings with large spans and/or weak rock are usually affected by breakdown or collapse processes that involve the detachment and fall of blocks and their accumulation on the cave floor. The resulting collapse breccias commonly display openwork texture and chaotic fabrics. The size and geometry of the blocks are largely controlled by the spacing and orientation of the discontinuity planes in the rock. Many caverns are related to the collapse of deeper dissolutional caves buried by breakdown deposits. Gravitational stresses tend to deflect around cavities creating a compression arch underlain by a tension zone that controls the development of cupola-shaped failure planes (Figs. 5.25A and 5.25B). Cavities may propagate upwards (stoping) by successive collapses controlled by arched unloading cracks, generating a cone-shaped breakdown pile on the cave floor, overlain by a collapse chimney with

Fig. 5.20C Shelfstones in Pas de Vallgornera Cave, Llucmajor, Mallorca (Published with permission of © Tony Merino 2015. All Rights Reserved)

a cupola-shaped roof. Another alternative is the downward migration of granular material from the cover or soft bedrock through dissolutional conduits. These feeders may supply detrital deposits into the cave and generate debris cones (Fig. 5.25C). This internal erosion process is responsible for the development of cover collapse sinkholes and cover suffosion sinkholes at the surface. Both collapse and internal erosion processes may be favoured by a change from phreatic into vadose conditions due to loss of buoyant support. Fresh breakdown deposits and debris cones in caves may be used to anticipate the location of future sinkholes.

5.26 Solution Notch

The bedrock surfaces of solution caves are commonly sculptured by dissolution features collectively called speleogens, in contrast to speleothems (mineral deposits). There is wide variety of speleogens. Here, two of the most common landforms that provide greater insight on cave development (speleogenesis) are described. Solution notches are horizontal furrows that mark the position of a persistent water level or the top of a pre-existing sediment fill. Their formation is related to more rapid wall recession at those levels due to more intense dissolution and often abrasion. Solution

Fig. 5.20D Associated crescentic shelfstones and rimstones in Pas de Vallgornera Cave, Mallorca. The shelfstones growing on the stalactites record changes in the water level of the pool (Published with permission of © Tony Merino 2015. All Rights Reserved)

notches in caves may be valuable geomorphic evidence for paleohydrogeological and sea-level reconstructions (Fig. 5.26).

5.27 Scallop

The speleogens are solutional concavities formed by turbulent water flow on bedrock surfaces. Their persistent asymmetry indicates the direction of the formative water flow, with the steeper side facing downflow. The morphometric characteristics of the scallops, mainly their length, may be used to estimate the velocity of the water flow responsible for their development. Their length ranges from a few millimetres to two metres; the larger the size, the lower the velocity. Scallops are best developed in massive, pure and homogeneous rocks. At very high velocities, abrasion by the transported particles precludes the formation of scallops (Fig. 5.27).

5.28 Tufa Deposits

The term tufa is commonly applied to highly porous carbonate deposits with abundant vegetation remains that form in ambient temperature freshwater environments. They may develop in a wide range of geomorphic and hydrologic settings, including fluvial channels; base-level, perched and artesian springs; lakes; and paludal environments. Calcium carbonate precipitation is essentially related to CO_2 loss

Fig. 5.21A Stalactites with branching helictites in Cueva de Cristal, Molinos, Iberian Chain, NE Spain (Published with permission of © Francisco Gutiérrez 2015. All Rights Reserved)

induced by biological activity (e.g. photosynthesis) and degassing in places with increased turbulence. It may be also related to supersaturation induced by water mixing (e.g. common ion effect). Calcium carbonate typically encrusts living macrophytes and microphytes forming autochthonous phytoherms. These growths, sometimes aided by log jams, may coalesce and develop barriers and cascades in fluvial systems, creating lakes or pools of calm water upstream (Fig. 5.28A). The destruction of phytoherms and other build-ups produce allochthonous detrital particles called phytoclasts and intraclasts. In flowing water, detrital particles may be progressively coated by multiple layers of carbonate forming coated grains or oncoids. Fine-grained micritic tufa and stromatolites mainly accumulate in lakes and pools with clam water. Tufa deposition in perched springs may form large prograding systems with curtains and overhangs. Tufa mounds form by deposition of carbonates around artesian springs (Fig. 5.28B). Similar accumulations but with different mineralogical composition (e.g. gypsum or halite) may be found in special conditions (Fig. 5.28C).

5.29 Travertine Deposits

The word travertine is considered by some authors as a synonym of tufa. Others contend that tufa is a weaker variety of travertine inadequate for construction. However, travertine is the preferred term for designating calcium carbonate deposited from thermal springs. Travertines mostly form from calcium bicarbonate hot waters with a high content of dissolved CO_2. Such thermal waters rapidly release CO_2 when they reach the surface because the partial pressure of this gas in the atmosphere is much lower. The removal of carbon dioxide may be accelerated by turbulence, especially at flow obstacles. In contrast with tufa deposits, travertines have very limited plant remains and biological activity does

Fig. 5.21B Straw stalactite in Sima del Portillo, Iberian Chain, NE Spain (Published with permission of © Juan Carlos Gordillo 2015. All Rights Reserved)

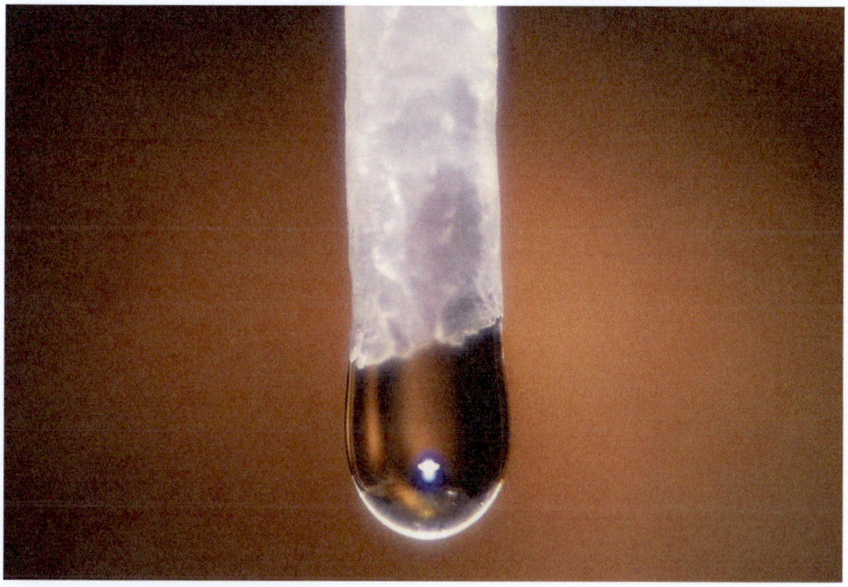

Fig. 5.21C Stalagmites with an apron of flowstone at the base in Lipska Cave, Montenegro. Note the broken stalagmite, which might be related to seismic activity (Published with permission of © Francisco Gutiérrez 2015. All Rights Reserved)

not play a major role in their deposition. They tend to be denser, less porous and more resistant. Laminated and well-bedded facies are common. Travertines show a wide range of morphologies conditioned by geomorphic and hydrological factors, including cascades, rimstones and pools, channel ridges and mounds. Channel ridges form by the supraelevation of channels due to rapid deposition on the beds and banks. Mounds form by precipitation around a spring vent in which water rises under pressure (Figs. 5.29A, 5.29B and 5.29C).

Fig. 5.21D Broken and welded column in the Peluda Cave, Atapuerca, Spain. Note the suspended tree roots in the ceiling of this shallow passage (Published with permission of © Francisco Gutiérrez 2015. All Rights Reserved)

Fig. 5.21E Translucent draperies associated with stalactites in Pas de Vallgornera, Llucmajor, Mallorca (Published with permission of © Tony Merino 2015. All Rights Reserved)

Fig. 5.21F Spheroidal and rod-like cave pearls in a shallow pool of Galeria del Tragus, Pas de Vallgornera Cave, Llucmajor, Mallorca (Published with permission of © Tony Merino 2015. All Rights Reserved)

Fig. 5.22 Flowstone with small rimstones and gours in Vjetrenica Cave, Popovo Polje, Bosnia–Herzegovina (Published with permission of © Francisco Gutiérrez 2015. All Rights Reserved)

Fig. 5.23A Helictites and popcorn in Sima de Val de Oton, Iberian Chain, Teruel (Published with permission of © Juan Carlos Gordillo 2015. All Rights Reserved)

Fig. 5.23B A shield with dripstones on its lower side, Ogle Cave, Carlsbad Caverns National Park, Eddy County, New Mexico (Published with permission of © Ira Sasowsky 2015. All Rights Reserved)

Fig. 5.24A Halite cave popcorn formed in a sector of Remolinos salt mine (Ebro Cenozoic Basin, NE Spain) with permanent air flow. The inclination of the stalactite is related to more rapid evaporation and precipitation on the windward side (Published with permission of © Francisco Gutiérrez 2015. All Rights Reserved)

Fig. 5.24B Cave coral in Sima de Val de Oton, Iberian Chain, NE Spain (Published with permission of © Juan Carlos Gordillo 2015. All Rights Reserved)

Fig. 5.25A Breakdown pile related to the upward stoping of a collapse cupola in Remolinos salt mine, Ebro Cenozoic Basin, NE Spain. Equivalent features form in natural caves (Published with permission of © Francisco Gutiérrez 2015. All Rights Reserved)

Fig. 5.25B Breakdown deposits in Obodska Cave, Montenegro. The individualization of the large blocks is controlled by stratification planes dipping to the right and cross joints (Published with permission of © Francisco Gutiérrez 2015. All Rights Reserved)

Fig. 5.25C Debris cone related to the downward migration of detrital particles through conduits (Peluda Cave, Atapuerca, Spain) (Published with permission of © Francisco Gutiérrez 2015. All Rights Reserved)

Fig. 5.26 Multiple horizontal solution notches developed of folded halite in Cova del Riu, Cardona salt diapir, Ebro Cenozoic Basin, NE Spain (Published with permission of © Ferrán Cardona 2015. All Rights Reserved)

Fig. 5.27 Scallops around 5 cm long in Wolf River Cave, Fentress County, Tennessee (Published with permission of © Ira Sasowsky 2015. All Rights Reserved)

Fig. 5.28A Chain of tufa barriers with associated waterfalls and pools in Plitvice Lakes National Park, Croatia (Published with permission of © Francisco Gutiérrez 2015. All Rights Reserved)

Fig. 5.28B Peculiar pole- and mound-like tufa deposits in Mono Lake, California. These built-ups form by precipitation of calcium carbonate at points where freshwater springs discharge under pressure into the lake bottom and mix with its alkaline water (Published with permission of © Francisco Gutiérrez 2015. All Rights Reserved)

Fig. 5.28C Unique tufa-like halite barriers in a stream that drains Jahane diapir in the Zagros Mountains, Iran (Published with permission of © Francisco Gutiérrez 2015. All Rights Reserved)

Fig. 5.29A Cascade of travertine deposits in Bighorn River, Thermopolis, Wyoming (Published with permission of © Francisco Gutiérrez 2015. All Rights Reserved)

Fig. 5.29B Cascade with rimstones and gours in Mammoth Hot Springs, Yellowstone National Park, Wyoming (Published with permission of © Francisco Gutiérrez 2015. All Rights Reserved)

Fig. 5.29C Travertine mound developed on the artificial Teppe Fountain built in 1909, Thermopolis, Wyoming (Published with permission of © Francisco Gutiérrez 2015. All Rights Reserved)

Granite Landforms

6

Granite is a common acidic plutonic rock whose major mineralogical components include quartz, alkali feldspar (K–, Na–feldspar), plagioclase (Na–Ca–feldspar) and mica. The petrological definition of granite refers to rocks in which quartz accounts for 20–60 % and plagioclase 10–65 % of the sum quartz + alkali feldspar + plagioclase. However, in the majority of the geomorphological works, the word granite is used in a broader sense to include other rocks that hold the same assemblage of landforms, like granodiorites, that are very common granitoid rocks with a high proportion of plagioclase. Outcrops of granitic rocks represent around 15 % of the continental area. They are particularly common in the tectonically stable shield areas (cratons) of the continental lithospheric plates and in recent orogenic areas affected by uplift and erosional unroofing.

Granitic landscapes display specific landforms determined by the characteristics of the rock type and their spatial variability. Granitic rocks are typically hard, homogeneous and crystalline rocks with a low primary porosity. The geometry of the granitic intrusive body may control the overall morphology of the landscape. Large discordant intrusions are called plutons and batholiths, the latter formed by several amalgamated plutons. Granitic plutons may be also surrounded by ring and radial dykes. Dykes are tabular and discordant intrusions related to the injection of magma along fractures. Laccoliths and sills are dome-shaped and tabular bodies roughly concordant with the host rock, respectively. The mineralogical composition of the rock influences chemical weathering, which is more intense on the less stable minerals. According to Goldich's weathering series, which has the reverse order to Bowen's crystallisation series, the minerals that formed at higher temperatures and pressures are less stable in meteoric environments and more susceptible to weathering. In the case of granitoids, quartz is the most stable mineral, whereas Ca-rich plagioclase, amphiboles and biotite are the less resistant to alteration. Hydrolysis of plagioclase and alkali feldspar is the main chemical weathering reaction resulting in the production of clay minerals (e.g. kaolinite). Slight weathering of minerals, especially along the crystal edges, reduces the intergranular strength favouring the disintegration of the rock. The textural characteristics of the rock, like the size of the constituent crystals, may have some control on the morphogenesis. The development of small-scale features may be favoured by a fine-grained aplitic texture or constrained by the presence of large crystals in rocks with pegmatitic or porphyritic texture.

Fractures are commonly a first-order controlling factor. They compartmentalise the rock mass into blocks and facilitate the penetration of meteoric water, favouring differential weathering and erosion. Relevant attributes of the fractures include spacing, relative orientation, width and shearing deformation in the case of faults. Granitic rocks commonly display a suborthogonal network of fractures. They may also display subhorizontal and typically discontinuous joins called pseudobedding. Moreover, erosional unloading may result in the development of dilation joints parallel to the topographic surface through a process called sheeting. Weathering preferentially acts along fractures and their intersections and progresses towards the inner parts of the compartments that they individualise. The rock is progressively transformed into a decomposed and friable material called grus that may surround rounded blocks of unweathered rock called corestones. Long-sustained alteration may produce deep weathering profiles. The boundary between the fresh bedrock and the overlying altered material, called the weathering front, typically has an irregular geometry related to differential weathering controlled by factors like the fracture density or petrological variation. The development of a number of granite landforms is attributed to the exhumation of irregular weathering fronts by differential erosion.

6.1 Inselberg

Inselberg is a German word that means "island hill". It refers to prominent steep-sided hills of resistant rock rising abruptly form an erosional plain (Fig. 6.1B). They are particularly frequent in granitoids and gneiss, but also form in other litholo-

F. Gutiérrez, M. Gutiérrez, *Landforms of the Earth*, DOI 10.1007/978-3-319-26947-4_6

gies like homogeneous sandstone formations (Fig. 6.1A). Inselbergs may rise hundreds of metres above the surrounding plains and show a broad morphological spectrum with three main end members. Domed inselbergs commonly developed in massive and poorly jointed rocks; castellated inselbergs, characterised by towers and pillars controlled by jointing; and boulder inselbergs, consisting of a relatively chaotic mass of boulders. The latter type can be derived from the degradation of the former ones. Inselbergs can be found in any morphoclimatic environment, although they are more common in tropical, arid and semiarid landscapes. The origin of these landforms may be related to different styles of landscape evolution (morphologic convergence or equifinality). Some authors envision the origin of inselbergs within a context of pediplanation, whereby the development of pediplains by progressive slope retreat results in the formation of isolated residuals. Other authors attribute the formation of inselbergs to deep differential weathering controlled by lithological and structural factors and the subsequent erosion of the regolith mantle.

6.2 Bornhardt

Bornhardts are dome-shaped, steep-sided hills mainly developed in resistant acid igneous rocks (e.g. granitoids, rhyolite). Their slopes are characterised by regular and convex profile on bare bedrock. Although there is some overlap with inselbergs, bornhardts do not necessarily occur as residuals in plains and have distinctive morphological and genetic features. The formation of these hills is related to deep structurally controlled weathering and the removal of the regolith overlying a highly irregular weathering front (rockhead). Bornhardts occur associated with compartments where the bedrock is less susceptible to weathering and erosion due to lower joint density and/or compositional differences. Concentric unloading joints (mega-exfoliation) develop on these massive and nearly isotropic crystalline rocks, which favour the development of dome-like shapes. The formation and dilation of these slope-parallel joints (sheeting) controls rock slides and rock falls that feed bouldery talus accumulations at the foot of the slopes (Figs. 6.2A and 6.2B).

6.3 Tor and Nubbin

Tors are medium-scale residual hills typically less than 25 m high, widely represented in granite terrains. They result from single or multiple cycles of deep differential weathering and stripping of the regolith mantling an irregular rockhead. The tors correspond to bedrock compartments less vulnerable to weathering mainly due to lower joint density. Some authors use the word tor to designate residuals with a wide range of morphologies. Others restrict the term tor or castle koppie to steep-sided castellated residuals. The latter archetypal tors are commonly controlled by an orthogonal joint system with two master vertical and subperpendicular joint sets, plus in some cases a subordinate set of subhorizontal joints (pseudobedding) (Fig. 6.3A). Tors at an advanced stage of decay may consist of a towering pile of rounded blocks and boulders. *Nubbin* refers to dome-shaped residual hills strewn with blocks or boulders (Fig. 6.3B). This surficial layer of boulders is attributed to the disintegration of the outer shell of the dome controlled by concentric sheet jointing (Fig. 6.3C).

6.4 Corestone and Granite Boulder

Subsurface weathering in well-jointed granite operates preferentially along the discontinuity planes and their intersections. The transformation of the hard bedrock into a decomposed and friable regolith (grus) progresses inwards from the joints towards the core of the compartments that they individualised. Corestones are the hard residual blocks surrounded by soft grus located in weathering mantles (Fig. 6.4A). Their maximum sizes are determined by the joint spacing. They tend to become progressively smaller and more rounded as weathering proceeds and may have concentric weathering layers (spheroidal weathering). Corestones may be exposed at the surface when the soft residual mass is removed selectively by erosion. The resulting boulders may be left in place or may suffer some transport. Granite boulders may also form by weathering and erosion processes guided by jointing at the surface. The boulders may reach more than 30 m in length, although the modal dimension is typically close to 1 m (Figs. 6.4B and 6.4C).

Fig. 6.1A Ayers Rock or Uluru in central Australia, a dome-shaped inselberg consisting of steeply dipping arkosic fine-grained conglomerate (Published with permission of © Francisco Gutiérrez 2015. All Rights Reserved)

Fig. 6.1B Spitzkoppe in the Namib Desert is among the highest inselbergs on Earth. This residual hill of complex morphology made up of Jurassic granite is nearly 700 m high and rises from an extremely flat plain cut-across Precambrian bedrock (Published with permission of © Francisco Gutiérrez 2015. All Rights Reserved)

Fig. 6.2A Granite bornhardts whose shapes are controlled by fractures and sheeting in Yosemite National Park, California. The highest hill is Mount Starr King (Published with permission of © Francisco Gutiérrez 2015. All Rights Reserved)

Fig. 6.2B The Half Dome in Yosemite National Park, California, is the remaining portion of a bornhardt partially removed by the excavation of a glacial trough. The steep wall on the margin of the valley is more than 1400 m high (Published with permission of © Francisco Gutiérrez 2015. All Rights Reserved)

Fig. 6.3A Granite tor ca. 8 m high with angular shape controlled by two vertical joint sets and horizontal pseudobedding (Karkonosze Mountains, SW Poland) (Published with permission of © Piotr Migon 2015. All Rights Reserved)

Fig. 6.3B Granite tor at an advanced state of degradation as revealed by the rounded and dislocated blocks (Manqueospese Castle, Avila, central Spain) (Published with permission of © Carles Roqué 2015. All Rights Reserved)

Fig. 6.3C Dome-shaped granite hill sparsely covered by detached rounded and angular boulders, Arabian Shield west of Hail, Saudi Arabia (Published with permission of © Francisco Gutiérrez 2015. All Rights Reserved)

Fig. 6.4A Weathered granite showing corestones surrounded by decomposed (grus) rock. Unweathered aplite dyke at the base of the excavation, La Vajol, Girona, Spanish Pyrenees (Published with permission of © Carles Roqué 2015. All Rights Reserved)

Fig. 6.4B Joint-controlled granite boulders with desert varnish, An Niayy, SE of Hail, Saudi Arabia (Published with permission of © Francisco Gutiérrez 2015. All Rights Reserved)

Fig. 6.4C A pair of nearly spherical balanced granite boulders in Devils Marbles (Karlu Karlu) Conservation Reserve, Northern Territory, Australia (Published with permission of © Scott Weatherson 2015. All Rights Reserved)

7 Weathering Landforms

Weathering refers to a large spectrum of processes that cause the breakdown and decomposition of materials at or near the surface. These alteration processes may involve fragmentation, chemical and mineralogical changes or even the incorporation of the main constituents of the material into meteoric waters as solutes. Most rocks have formed under different conditions to those found in the near-surface environments where they exist now. Weathering essentially represents changes in the rocks towards equilibrium conditions in the surface and near-surface environments. Weathering plays an instrumental role in numerous earth-surface processes, the development of specific landforms and the long-term evolution of landscapes. It prepares materials for subsequent erosion, produces regoliths and is an essential process in soil development (pedogenesis). In some settings, such as karst environments, weathering by dissolution is the dominant morphogenetic process. Weathering processes are typically classified into physical, chemical and biological, although there are frequently some overlaps and interrelationships. Moreover, a significant number of similar weathering landforms may be produced by different processes (morphological convergence or equifinality).

Physical weathering involves the fragmentation or disintegration of materials by near-surface stresses without any chemical or mineralogical change. The forces can be produced by numerous processes and commonly involve differential volume changes in minerals, particles and/or voids. Salt weathering is related to pressures generated by the crystallisation and hydration of salts in intergranular spaces and cracks. Oversaturation conditions may be reached by evaporation, temperature changes or even the mixing of waters with different chemical compositions. Temperature changes may produce expansion and contraction cycles fragmenting the rock by a process of insolation weathering. These volume changes may occur differentially on minerals with different colours that absorb a variable amount of heat and between the outer and inner parts of the rocks. Similarly, wetting and drying cycles also cause volume changes; they are particularly intense in materials with swelling clays. Frost shattering is related to the expansion of water during freezing and most importantly the migration of water towards zones of ice segregation. Unloading in overconsolidated massive rocks produces stress release joints that tend to be parallel to the topographic surface, a process called sheeting.

Chemical weathering involves changes in the chemical and/or mineralogical composition of the materials. The relative importance of the different chemical weathering processes (e.g. dissolution, hydrolysis, hydration, redox reactions) largely depends on the composition of the rocks. Dissolution entails the transfer of atoms or molecules from the mineral surface to a solution. It may cause the complete removal of huge volumes of soluble rocks (karstification) by water dissolving evaporites or carbonates. It may also cause a substantial reduction in the intergranular cohesion by the dissolution of cements, a process that favours rock disintegration. Hydrolysis involves the replacement of metal cations in silicate minerals by hydrogen cations and the combination of the released cations with hydroxyl anions. These reactions lead to the formation of secondary clay minerals. Hydration is the transformation of a mineral into another by the incorporation of water in the mineral lattice (e.g. hydration of anhydrite into gypsum). Oxidation mainly affects minerals containing iron and results in the formation of iron oxides and the release of electrons.

Plants, lichens, algae and bacteria may significantly contribute to weathering through physical and chemical mechanism (biophysical and biochemical weathering). Growing roots produce and dilate fractures in rocks (root wedging) and cause bioturbation in unconsolidated deposits. The hyphae of lichens produce microcracks and reduce intergranular strength. Lichens and algae may experience significant volume changes controlled by moisture variation. Furthermore, these organisms may produce organic acids increasing the aggressiveness of the water. Acidity is also increased (lower pH) by the decay of organic matter and the consequent release of CO_2.

F. Gutiérrez, M. Gutiérrez, *Landforms of the Earth*, DOI 10.1007/978-3-319-26947-4_7

Weathering progressively causes the alteration of rocks, transforming them into a residual material usually called regolith. The boundary between the fresh bedrock and the overlying altered material is called the weathering front, which migrates downwards as weathering proceeds, and may be more than 100 m deep. This boundary may be planar or irregular, net or gradational. Weathering rates depend on a complex set of intrinsic and extrinsic factors, notably lithology and climate. Chemical weathering processes like hydrolysis and dissolution may reach very high rates in warm tropical environments with high precipitation. Increases in temperature accelerate the kinetics of numerous chemical reactions. Water is the main agent in most weathering processes responsible for the leaching or lixiviation of soluble components. In contrast, frost shattering is more intense in cold regions with a high annual number of freeze–thaw cycles.

Past and active weathering products and processes have relevant applied implications. Weathering is an essential process in the development and renewal of soils, one of the crucial requirements on Earth for human life. Some weathering products, such as laterites, are valuable mineral resources. Weathering may produce severe damage to human structures, including historical buildings, through stone decay. It also plays an important role in the carbon cycle influencing the concentration of atmospheric CO_2 that is responsible for the greenhouse effect and the consequent global warming. As this happens, the increase in the concentration of atmospheric CO_2 is expected to cause an intensification of the dissolution processes affecting carbonate rocks.

7.1 Tafoni and Alveoli

Cavernous weathering refers to differential weathering processes that lead to the formation of hollows on steep rock slopes and boulder faces. The term tafoni (singular: tafone) is used to designate decimetre- to metre-scale niches, whereas centimetre-sized hollows are termed alveoli (singular: alveole) (Figs. 7.1A, 7.1B, 7.1C, and 7.1D). These features form in a wide variety of lithologies and environmental conditions, although they are particularly frequent in some hard rock types, such as sandstones or granitoids, as well as in coastal and arid regions. Tafoni typically display subspherical or ellipsoidal geometry, with bowed entrances, overhanging visors, concave inner walls and bottoms with a thin veneer of weathering-derived deposits. Alveoli tend to form a mesh of closely spaced hollows that resemble a honeycomb (honeycomb weathering). These hollows become progressively deeper and may coalesce forming larger niches that eventually evolve into tafoni. The main mechanisms involved in the differential recession of the rock faces are granular disintegration and flaking, which may be related to a combination of weathering processes (e.g. salt weathering, chemical alteration, wetting and drying, frost action). Salt weathering related to stresses induced by the crystallisation and hydration of salts in intergranular spaces and cracks is considered one of the main weathering processes. This notion is supported by the common occurrence of saline efflorescences in the walls and deposits of the hollows and the higher frequency of tafoni and alveoli in salt-rich environments. The evacuation of the weathering products that accumulates in the hollows is achieved by wind action and animals. The tafoni that develop in rock fins may eventually create holes, and its growth may lead to the development of arches (Figs. 7.1E and 7.1F). The holes under arches typically grow upwards through rock falls controlled by curved stress release cracks. Gravitational stresses deviate around the voids generating an arched tension zone above the roof of the cavity where the joints form. The occurrence of well-developed cavernous weathering in buildings illustrates the detrimental impact of cavernous weathering on human structures and the relatively high rate at which they form (Fig. 7.1D).

7.2 Gnamma

Gnammas, also known as weathering pits, are small closed depressions formed by differential weathering on flat or gently sloping rock surfaces. They are the equivalent of solution pans or kamenitzas generated by solution in non-karst rocks. Gnammas typically have a decimetre- or metre-scale diameter, but may reach more than 10 m across. They commonly display a flat floor that may be covered by residual material. The edges are typically scarped and may show an overhanging profile, especially in case-hardened rock surfaces, and may be incised by overflow channels. Similarly to tafoni and alveoli, they develop in a wide range of lithologies and morphoclimatic environments. The physical, chemical and biological weathering processes are also similar. Probably the main differences are temporary ponding, higher water infiltration through their floor and the evacuation of the detrital sediments by water overflow in addition to wind deflation (Figs. 7.2A, 7.2B, 7.2C, and 7.2D).

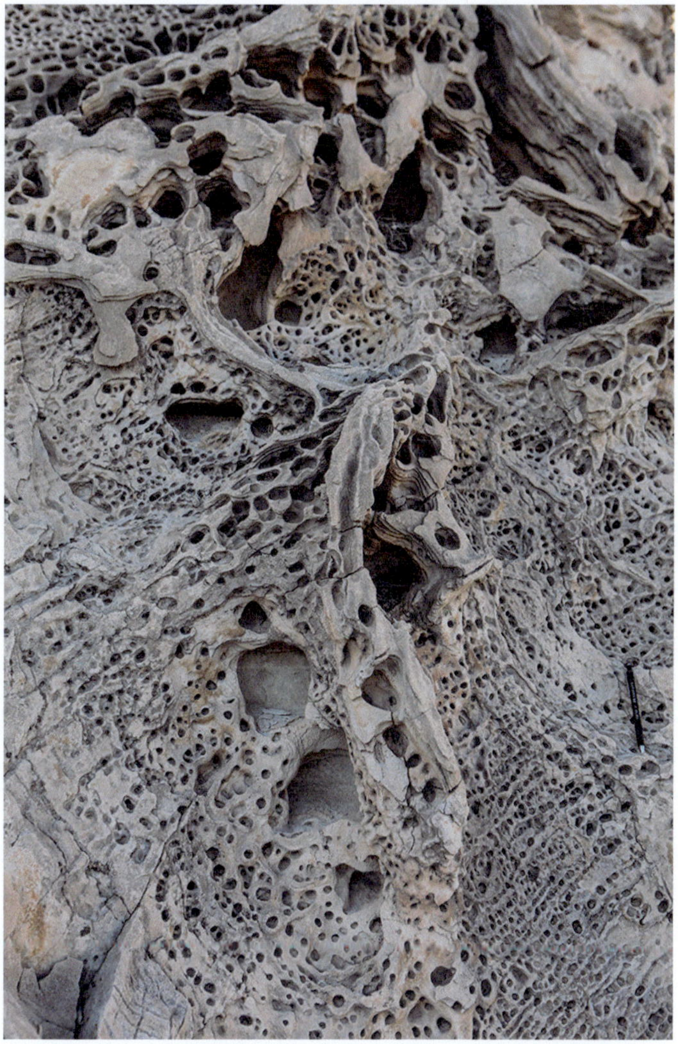

Fig. 7.1A Intricate mesh of alveoli and tafoni in a steep slope on cross-bedded and jointed sandstone next to a salt extrusion in the Zagros Mountains, Bastak area, Iran (Published with permission of © Francisco Gutiérrez 2015. All Rights Reserved)

7.3 Flared Slope

These are rock slopes with an overhang and a concavity at the base. These notches, typically a few metres high, are especially frequent in granitoids, but may also form in a wide variety of rock types. The flares are related to more effective subsurface weathering below the foot of rock slopes, where the regolith receives a large amount of aggressive runoff water from the rock outcrop situated above. During long periods of landscape stability, the regolith–bedrock contact associated with the base of the slopes becomes progressively steeper as weathering progresses downwards and backwards. The flare is eventually exposed by erosion. Rock slopes may display flares at different elevations, which may record alternating phases of landscape stability, dominated by weathering and regolith development, and periods of instability during which the regolith is eroded (Figs. 7.3A and 7.3B).

7.4 Spalling and Flaking

Spalling, also termed flaking, refers to the detachment of weathered planar fragments from the rock surface. The spalls or flakes are typically centimetre to decimetre in scale and their width is generally less than 5 cm. The outward separation of a continuous shell or discontinuous flakes in rock surfaces is mainly related to differential stresses associated with weathering-related volume changes. The expansion of the outer zone may be induced by precipitation of secondary minerals, including salts, growth of ice crystals, thermal expansion or lichen growth. The spalling process is frequently observed in case-hardened rock surfaces, characterised by an outer indurated shell related to the precipitation of secondary minerals, which may include exotic material not found in the host rock (Figs. 7.4A and 7.4B).

Fig. 7.1B Cavernous weathering developed on the arkosic fine-grained conglomerate of Ayers Rock (Uluru), central Australia (Published with permission of © Francisco Gutiérrez 2015. All Rights Reserved)

Fig. 7.1C Sandstone slopes riddled by cavernous weathering in Petra, Jordan. Note case-hardened surfaces on the background and steps carved by the Nabateans on the left (Published with permission of © Francisco Gutiérrez 2015. All Rights Reserved)

Fig. 7.1D Stone blocks in a historical building in Santillana del Mar, northern Spain, showing a markedly different degree of weathering. Well-developed alveoli in poorly cemented sandstone block (Published with permission of © Francisco Gutiérrez 2015. All Rights Reserved)

Fig. 7.1E Delicate Arch hollowed in an isolated fin of Entrada Sandstone, Arches National Park, Utah (Published with permission of © Francisco Gutiérrez 2015. All Rights Reserved)

Fig. 7.1F Turret arch and a hole developed in a rock fin consisting of Carmel Formation overlain by Entrada Sandstone, Arches National Park, Utah (Published with permission of © Francisco Gutiérrez 2015. All Rights Reserved)

Fig. 7.2A Water-filled gnammas at the top of Ayers Rock, central Australia. The Olgas inselberg protruding over a vast planation surface in the background (Published with permission of © Francisco Gutiérrez 2015. All Rights Reserved)

Fig. 7.2B Metre-scale weathering pits partially filled by aeolian sand on a horizontal surface underlain by cross-bedded sandstone, Hail area, Saudi Arabia (Published with permission of © Francisco Gutiérrez 2015. All Rights Reserved)

Fig. 7.2C Large gnamma developed in the Jurassic-age Entrada Sandstone, Arches National Park, Utah (Published with permission of © Francisco Gutiérrez 2015. All Rights Reserved)

Fig. 7.2D Gnamma with overflow channel developed on Jurassic granite. Spitzkoppe inselberg in the background, Namibia (Published with permission of © Francisco Gutiérrez 2015. All Rights Reserved)

Fig. 7.3A Wave Rock at Hyden Rock, western Australia, is the classical example of flared slope developed in granitoids (Published with permission of © Keith Smith 2015. All Rights Reserved)

Fig. 7.3B Flared rock at the base of a rock slope consisting of steeply dipping arkosic sandstones and fine-grained conglomerates, Ayers Rock, central Australia (Published with permission of © Francisco Gutiérrez 2015. All Rights Reserved)

Fig. 7.4A Flakes ready for erosion on a rock slope developed on sandstones and conglomerates, Ayers Rock, central Australia (Published with permission of © Francisco Gutiérrez 2015. All Rights Reserved)

Fig. 7.4B Spalling in a marly limestone bed in Belsué Reservoir, Spanish Pyrenees (Published with permission of © Francisco Gutiérrez 2015. All Rights Reserved)

7.5 Spheroidal Weathering

This term refers to the progressive alteration of boulders or joint-bounded rock blocks through the development of a weathering cortex with concentric layering. This process may operate both at the surface and in the subsurface and is particularly well developed in granitoids. In the case of a well-jointed rock mass, weathering tends to preferentially operate at the vertexes, edges and planes of the rock compartments, in decreasing order of intensity. This joint-controlled alteration typically leads to the development of rounded residual blocks (corestones) with a shell of concentric layers that may be surrounded by a decomposed regolith. Weathering of boulders exposed at the surface may also advance inwards through the formation of concentric spalls. The detachment and erosion of the spalls favours the inward progression of the weathering processes and the reduction of the boulder size (Figs. 7.5A and 7.5B).

7.6 Frost Shattering

Frost shattering, also known as frost weathering, congelifraction, gelifraction or gelivation, refers to various rock breakdown processes related to freezing of water in pores and discontinuity planes. Frost shattering is a fundamental morphogenetic process in periglacial environments and typically produces angular clasts. Different stresses associated with the freezing process may be involved in the fragmentation or disintegration of the rock: (1) Volume expansion of water during freezing. (2) Suction tension related to water that migrates towards freezing zones and fronts, which typically progress from the surface towards inner and deeper parts of the rock mass. (3) Hydrostatic pressure in cracks related to water trapped by the sealing freezing front. The effectiveness of frost shattering depends on climatic factors, mainly freeze–thaw cycles, and the characteristics of the rock (e.g. density and width of discontinuity planes, porosity) (Fig. 7.6).

Fig. 7.5A Jointed granite affected by spheroidal weathering, Brandberg, Namibia (Published with permission of © Francisco Gutiérrez 2015. All Rights Reserved)

Fig. 7.5B Spheroidal weathering in jointed granite coated by desert varnish, An Niayy, Hail region, Saudi Arabia (Published with permission of © Francisco Gutiérrez 2015. All Rights Reserved)

Fig. 7.6 Boulder on a Little Ice Age frontal moraine of Fjallsjokull Glacier, southern Iceland, transformed into a pile of angular clasts mainly by frost shattering (Published with permission of © Francisco Gutiérrez 2015. All Rights Reserved)

7.7 Sheeting

Sheeting refers to the development of subparallel joints in rock massifs concordant with the topographic surface. The resulting sheets typically have a decimetre- to metre-scale thickness and are frequently curved. These secondary fractures are especially well developed in granitoids, but may also form in a wide range of igneous, metamorphic and sedimentary rocks. According to the most widely accepted interpretation, sheeting is related to near-surface stress release (unloading). This notion involves the formation of surface-parallel joints with a dilational vector perpendicular to the ground surface, resulting in a volume increase. The spacing of the joints tends to increase with depth as the confining stress increases. Other authors contend that sheeting in igneous rocks may form during cooling of the intrusive body. Implicitly, the exogenetic interpretation indicates that sheeting is controlled by the topography, whereas in the endogenetic model, the topography is controlled by pre-existing joints formed deep in the crust. Sheeting significantly contributes to reduce the mechanical strength of the rock mass producing planes highly favourable for the development of shallow rock slides (Figs. 7.7A, 7.7B, and 7.7C).

7.8 Defile

As in karst rocks, weathering along subvertical joints results in the development of defiles. The equivalent dissolution features in karst rocks are called grikes. Initially, these structural discontinuities may open by dilation related to unloading (stress release), which is typically more pronounced close to scarps where the rock mass has experienced the loss of lateral confining pressure. Water infiltration and subsurface flow tend to concentrate along these secondary porosity features, preferentially dissolving some mineralogical components (e.g. calcite, dolomite, quartz) of the rock at the joint faces. Arenization is considered the main process involved in weathering of quartzose sandstones. Slow chemical dissolution of quartz, especially at intergranular contacts, frees the grains and prepares them for physical weathering and erosion. These defiles tend to display subparallel walls, in contrast to the commonly downward tapering grikes in more soluble limestone rocks, where the aggressiveness of the water is rapidly reduced. Long-sustained widening of joints may result in the development of corridor-like defiles several metres wide. In areas with intersecting joint systems, these processes may produce a maze of corridors with intervening towers forming dramatic ruiniform landscapes (Figs. 7.8A and 7.8B).

Fig. 7.7A Granitoids affected by closely spaced joints concordant with the steep slope, Yosemite National Park, California (Published with permission of © Francisco Gutiérrez 2015. All Rights Reserved)

Fig. 7.7B Steep slope with a sheet fracture cross-cutting bedding planes in the fine-grained conglomerates of Ayers Rock, central Australia (Published with permission of © Francisco Gutiérrez 2015. All Rights Reserved)

Fig. 7.7C Surface-parallel unloading joint in conglomerates at the Olgas, central Australia (Published with permission of © Francisco Gutiérrez 2015. All Rights Reserved)

Fig. 7.8A Defiles in sandstone controlled by an orthogonal system of subvertical joints, west of Hail, northern Saudi Arabia (Published with permission of © Francisco Gutiérrez 2015. All Rights Reserved)

Fig. 7.8B Defile controlled by vertical joint in sandstone. On the right, the Monastery, a temple carved by the Nabateans on the first century AD in Petra, Jordan (Published with permission of © Francisco Gutiérrez 2015. All Rights Reserved)

7.9 Gypsum Tumuli

Gypsum tumuli are hollow subcircular domes generated by the expansion and bulging of weathering crusts in bare gypsum outcrops. These blister-like features are typically several decimetres in diameter and in height, and the upwarped layer may be as thick as 50 cm. Gypsum tumuli up to 15 m across and 1 m high have been documented in outcrops of macrocrystalline gypsum. The origin of these domes has been attributed to volume increase and buckling in the outer weathering zone by gypsum dissolution and reprecipitation. Their formation is favoured by semiarid conditions and a high number of wetting and drying cycles. Infiltration water dissolves gypsum in the upper part of the rock mass. Subsequently, evaporation and the associated capillary rise of the interstitial water lead to the upward migration of the solution, its progressive saturation and gypsum reprecipitation in the outer zone of the rock mass. The volume increase caused by secondary gypsum precipitation, together with crystallisation pressure, results in the detachment and bulging of a gypsum crust. As the domes grow, concentric and radial cracks may develop and eventually the top of the blister collapses producing a crater-like feature (Figs. 7.9A and 7.9B).

Fig. 7.9A Collapsed gypsum tumuli in the central sector of the Ebro Cenozoic Basin, NE Spain (Published with permission of © Francisco Gutiérrez 2015. All Rights Reserved)

Fig. 7.9B Collapsed gypsum tumuli in Sebkha El Melah, Tunisia (Published with permission of © Mateo Gutiérrez 2015. All Rights Reserved)

8 Slope Movements

Slope movements are also collectively referred to as mass movements or landslides. Some authors contend that the term landslide should be restricted to slope movements with well-defined failure surfaces or basal shear zones. Slope movements can be defined as the outward and downward displacement of slope-forming material under the influence of gravity. They have a very widespread distribution, since unstable landslide-prone slopes occur almost everywhere. Landslides are major agents of topographic change and landscape evolution. They are one of the main processes involved in the destruction of relief and the delivery of sediment to fluvial systems. Large slope movements may block rivers creating landslide-dammed lakes, whose sudden outburst may lead to extreme floods and the mobilisation of huge volumes of sediment. Landslides can also control the growth of volcanic edifices, which may be limited by flank collapses that can drastically change the behaviour of volcanic eruptions. Landslides are very common in the seafloor, especially on the continental slope, and include the largest mass movements on Earth, reaching volumes of the order of 10^{13} m^3. Some of the main triggers for terrestrial landslides include severe rainfall events, seismic shaking, undercutting related to fluvial or marine erosion and a wide range of human alterations, including excavation and deforestation. Landslides are particularly abundant in tectonically active mountain belts, where some natural triggers have a significant impact: precipitation that is enhanced by the orographic barrier effect, large earthquakes and fluvial incision stimulated by tectonic uplift.

Slope movements can be classified using two descriptors. The first one identifies the material affected by the displacement (rock, debris and earth). Rock is hard lithified material, and debris and earth are used for unconsolidated soils (i.e. detrital or residual deposits) with a high proportion of coarse clasts or mostly composed of fine-grained particles, respectively. Other more specific terms like mud or sand may be used instead of earth. The second term indicates the movement mechanism: slide, lateral spread, flow, topple and fall. This order roughly reflects a gradation from movements in which the mass moves as a coherent body, to incoherent disintegrated masses, and to lose particles that displace independently. Frequently, various types of movement occur simultaneously in different parts of a landslide (composite landslides), or the movement style and/or material type may change as the mass moves and modifies its rheology (complex movements). Composite and complex movements can be described using combinations of the different terms (e.g. earth slide flow). Some special slope movements are named using specific terms that do not follow this classification scheme (e.g. sackung, solifluction).

Landslides cause significant economic losses in most countries, and rapid slope movements have the potential to cause the large-scale loss of human life. A dramatic example is the earthquake-triggered Huascarán rock avalanche in the Peruvian Andes, which killed over 66,000 people. Landslides were responsible for over 10,000 deaths and left 2.5 million people homeless over the decade 2001–2010. The damage may be related to the slope movements itself, or to secondary processes, like floods produced by the outburst of landslide dams or impulse water waves (tsunamis) created by large and rapid landslides in water bodies. The 1963 Vaiont landslide, Italy, generated an impulse water wave in the Vaiont Reservoir and a flood that killed around 2600 people. During the second half of the twentieth century, there has been a rapid increase in the economic losses related to landslides. It is likely that this raising trend will continue for forthcoming decades, as vulnerable elements (population and property) keep on increasing and human activity has greater detrimental impacts on slope stability.

8.1 Translational Slide

A slide occurs when a coherent mass of rock or soil moves along discrete failure planes or narrow shear zones. Translational slides move on planar slip surfaces following an approximately rectilinear displacement. The failure planes are commonly controlled by pre-existing structural or

F. Gutiérrez, M. Gutiérrez, *Landforms of the Earth*, DOI 10.1007/978-3-319-26947-4_8

stratigraphic discontinuities, such as faults, joints, bedding or the contact between bedrock and weak cover deposits (colluvium, residuum). Discontinuities with a strike approximately parallel to the orientation of the slope and with an outward dip close to or above the angle of friction are particularly favourable for the development of translational failures (Figs. 8.1A, 8.1B and 8.1C). Rock slides may displace as coherent blocks with limited internal deformation. On steep and high slopes, rock translational slides may attain sufficient velocity to progressively lose coherence, disintegrate and transform into extremely rapid rock avalanches (sturzstroms) (Fig. 8.1B). Failures controlled by the basal contact of colluvium or residual deposits produce debris slides with a shallow sliding plane commonly parallel ground surface (Fig. 8.1D). These soil slides, frequently triggered by pore water pressure rises related to severe rainfall events, tend to promptly disintegrate and transform into debris flows. This type of landslides may also form in permafrost areas, where the active layer lying on impervious and stiff frozen ground experiences a significant decrease in strength by thawing (Fig. 8.1E). Wedge failures are translational slides formed by displacement along a wedge-shaped slip surface that results from the intersection of two discontinuity planes. The outward dipping intersection line defines the displacement trajectory of the moving mass. These slope movements frequently occur on rock cliffs and transform into rock falls (Fig. 8.1F).

8.2 Rotational Slide

Rotational slides move along curved and concave-up slip surfaces. The landslide body experiences an inward rotation about an axis perpendicular to the displacement direction and situated above the slope. Rotational slides typically move as coherent blocks with limited internal disruption. They are particularly common in homogeneous argillaceous soils and rocks and in stratigraphic successions consisting of resistant rocks underlain by weaker clayey units (Fig. 8.2A). Backtilting of the topographic surface in the upper part of the landslide produces conspicuous benches and enclosed depressions that may contain ponds or lakes (Figs. 8.2B and 8.2C). The removal of the lateral confining pressure (debuttressing) in the headscarp of the landslide may favour the development of successive multiple retrogressive failures. Rotational slides, in addition to backward rotation, commonly also have a translational displacement component. The latter may create a void space above the listric failure plane in the upper part of the landslide, which may lead to the development of antithetic normal faults and a keystone graben. Compression in the lower part of the landslide tends to produce a bulge. This zone may lose coherence and transform into a secondary flow-type movement.

8.3 Lateral Spread

Lateral spreading is the lateral extension of a rock mass or cohesive soil in a slope induced by the deformation of a soft underlying material, without the development of discrete, well-defined failure surfaces. Rock lateral spreads typically develop on slopes consisting of brittle rocks underlain by soft formations (e.g. mudstone, salt). The laterally unconfined soft rock under the load of the overlying rigid slab may expand outwards, eventually producing a bulge. This extension is accommodated in the overlying caprock through the retrogressive development of tensile fractures subperpendicular to the displacement direction and the dragging apart of the intervening blocks (Fig. 8.3A). The separation of the blocks may occur through pre-existing tectonic fractures or newly formed gravitational joints. As spreading progresses, a horst and graben morphostructure and deep extensional gashes may develop (Figs. 8.3B and 8.3C). The lateral extension of the soft rock unit is accompanied by vertical contraction, leading to subsidence of the overlying rigid blocks. Differential subsidence, which commonly increases towards the free face of the slope, may lead to the downward flexure of the rock slab and the outward toppling of the decomposed blocks. Rock lateral spreads in continental areas may reach volumes of the order of tens of cubic kilometres and are characterised by very low displacement rates. Soil lateral spreads commonly occur on cohesive soils underlain by weak fine-grained deposits affected by liquefaction or plastic deformation. The blocks of the cohesive material may undergo, in addition to outward translation, subsidence and rotation, and the whole mass typically transforms into a flow-type movement. This type of slope movements is characteristic of unconsolidated sensitive clays (quick clays) with thixotropic behaviour that undergo a sharp strength reduction by seismic shaking. Landslides in quick clays may from on low-angle slopes, reach high volumes and expand rapidly by retrogression.

8.4 Flow

Flow refers to a broad group of slope movements in which the mobilised material undergoes continuous deformation without the development of persistent and well-defined failure planes. Two main flow types can be differentiated from a rheological perspective: cohesive flows in which deformation is accommodated by visco-plastic deformation of a clayey material and cohesionless flows of loose particles that may interact by collisions with the resulting dispersive forces and comminution. The latter group includes the grain flows that occur in the lee side of eolian dunes and rock avalanches, addressed in a different entry. Debris flows are the most

Fig. 8.1A The 1963 Vaiont Reservoir translational slide developed on a limestone succession with interbedded marls (Italian Alps). The sliding surface was controlled by thin clayey beds highly dipping towards the valley. This catastrophic landslide, with an estimated volume of around 270 million m^3, created an impulse water wave that overtop the dam (lower right), resulting in a flood in the Piave valley (right side of the image) that killed around 2000 people (Published with permission of © Virtualgeo 2015. All Rights Reserved)

frequent flow-type slope movements. They are wet and muddy slurries of unconsolidated material that move downslope as a viscous flow. These are high-density and high-competence flows that are able to transport metre-sized boulders; upward buoyant forces acting on particles are much higher than in water flows. Debris flows can be highly mobile, reaching extremely high velocities and runouts. Flows with an evident lack of coarse particles can be described as mud flows. Debris flows frequently result from the disintegration of shallow soil slides. They can also derive from water flows in channels by a progressive increase in solid load concentration, changing into hyperconcentrated flows and then into debris flows. Debris flows that originate in hillslopes may travel as unconfined flows or confined along pre-existing channels. They frequently accumulate lateral ridges of debris along the channel edges called levees (Fig. 8.4A). These sediment gravity flows accumulate where the gradient flattens, typically forming lobe-shaped accumulations with relatively steep fronts (Figs. 8.4B and 8.4C). After deposition the debris flow surface may settle due to water loss (Fig. 8.4D). Debris flow facies are commonly unstratified, poorly sorted, with chaotic fabrics and may show reverse grading, with larger clasts at the top. The term lahar is used for debris flows develop in volcanoes, where loose fine-grained pyroclasts are abundant. The water responsible for the initiation of lahars may have various sources, including snow or ice melting, rainfall or crater lakes (Fig. 8.4E). The term earth flow is commonly restricted to drier, stiffer and slower flows with significant deformation at the boundaries. These flows are frequently derived from slides of argillaceous rocks that experience a significant reduction in strength and coherence loss after limited displacement (Fig. 8.4F).

8.5 Topple

Toppling is the forward rotation of rock or soil about an axis situated below the centre of gravity of the moving mass. Topples mainly develop on cliffs and steep slopes and involve discrete columns or slablike bodies separated by vertical or steeply dipping discontinuity planes (e.g. joints, unloading

Fig. 8.1B The 1925 Lower Gros Ventre translational slide, Wyoming, developed on a dip slope underlain by sandstone, limestone and shale dipping around 20° towards the valley. This rapid 40 million m^3 movement transformed into a rock avalanche that dammed the river and created a lake (Published with permission of © Francisco Gutiérrez 2015. All Rights Reserved)

cracks, bedding) (Figs. 8.5A, 8.5B and 8.5C). As forward rotation progresses, the block loses its basal support and the topple gives way to a fall. Most commonly, toppling affects blocks of rock or soil in a scarp compartmentalised by widely spaced joints. A less frequent mode of toppling is the outward flexing of layers or stratiform bodies bounded by closely spaced and steeply dipping discontinuities (flexural toppling). Topples frequently enlarge retrogressively due to progressive loss of lateral confining pressure (Fig. 8.5B). A wide variety of mechanisms may be involved in toppling including plastic deformation of clayey units underlying the blocks, undermining, ice wedging, dissolution-induced subsidence or shearing forces at the top of the blocks by a moving cover, among others.

8.6 Fall

Falls are preceded by the separation of a mass of rock or soil from a steep slope, until it is detached and starts to descend. The dropping mass may undergo three sequential movement types with distinct trajectories. Vertical free fall solely occurs on very steep cliffs. The blocks that impact on the slope surface may initially bounce through semi-elastic rebounds and subsequently roll as they lose momentum. Bouncing boulders may create impact craters on the ground. Rock falls commonly break up into progressively smaller particles (comminution) as a result of collisions among blocks and with the ground surface. Larger particles tend to travel longer distances (run out). The velocity of free-falling particles depends on the drop height, regardless of particle size, but the momentum of bouncing and rolling particles is proportional to their size (Fig. 8.6A). Rock falls may produce continuous accumulations at the foot of rock cliffs called talus slopes and talus cones (Fig. 8.6B). These deposits typically show longitudinal sorting with larger particles in the distal sectors. Rock falls are commonly preconditioned by preexisting discontinuities, whose spacing and orientation controls the size of the blocks. Some of the most frequent triggers involved in the development of falls include severe rainfall events, frost shattering, undermining, earthquakes, root wedging and permafrost thawing (Figs. 8.6A, 8.6B, 8.6C and 8.6D).

Fig. 8.1C Active translational slide developed on Paleozoic slates with a general dip towards the valley. A significant part of the landslide volume is submerged in the Lanuza Reservoir, Tena Valley, Spanish Pyrenees (Published with permission of © Francisco Gutiérrez 2015. All Rights Reserved)

Fig. 8.1D Scars of shallow soil slides in which the failure surface was controlled by the bedrock–colluvium contact, Vik area, southern Iceland (Published with permission of © Francisco Gutiérrez 2015. All Rights Reserved)

Fig. 8.1E Shallow soil slide transformed into an earth flow in a permafrost area of Mumien, central Spitsbergen. The landslide developed on the seasonally thawed active layer overlying the permafrost table (Published with permission of © Zbigniew Zwoliński 2015. All Rights Reserved)

8.7 Rock Avalanche

Rock avalanches, also named sturzstroms, are extremely rapid and mobile flows of loose blocks that typically derive from the disintegration of large rock slides or rock falls formed on steep and high slopes. A striking peculiarity of these turbulent flows is their long runout, mainly attributed to the rapid fragmentation of the rock mass into progressively smaller blocks (comminution) that experience continuous collisions. This process involves a rapid cohesion and strength loss and a momentum exchange between colliding particles that result in dispersive stress that reduce basal friction. Other mechanisms that may contribute to the high mobility of these slope movements include lubrication by trapped and compressed air, or the entrainment of water-rich surface deposits along their path, which contribute to increase their fluidity and volume. Rock avalanches are typically accompanied by a cloud of dust related to the brecciation process and may run up obstacles hundreds of metres high. The accumulations are fan-, lobate- or tongue-shaped debris sheets with hummocky surfaces, longitudinal and transverse ridges and furrows and a steep and sharp distal edge. The deposits are typically chaotic, unstratified, tightly packed and poorly sorted mega-breccias. The angular debris generally shows conspicuous reverse grading with an armour of large blocks underlain by clasts with a high proportion of fines related to the brecciation process. Rock avalanches typically occur in tectonically active mountain belts and volcanoes, and common triggers include heavy rainfall events, large earthquakes, volcanic eruptions or slope undercutting (Figs. 8.7A and 8.7B).

8.8 Sackung

Sackung generally is used to designate a type of gravitational spreading in rock slopes that is accommodated through the development of gravitational normal faults dipping into the slopes These non-tectonic surface ruptures, generally parallel to the ridge crests and contour lines, produce uphill-facing scarps (also called antislope scarps, counter scarps and anti-

Fig. 8.1F Wedge failure in a limestone quarry controlled by two conjugate joints that cross-cut bedding, Ricla, Iberian Chain, NE Spain (Published with permission of © Francisco Gutiérrez 2015. All Rights Reserved)

thetic scarps) and linear depressions lying at their foot (Figs. 8.8A and 8.8B). The scarps in the ridge tops may create double ridges and ridge-top depressions (Fig. 8.8C). Slopes affected by this deep-seated gravitational deformation commonly display a bulge in its lower part. The uphill-facing scarps are typically hundreds of metres long and a few metres high and generally display higher height-to-length ratios (aspect ratio) than tectonic fault scarps. The troughs associated with the antislope scarps may defeat drainages and contain enclosed depressions that behave as sediments traps. Sackung features have been described in a wide variety of geological environments, but they seem to be particularly frequent in glaciated and seismically active mountain areas. Trenching and geodetic investigations indicate that sackung scarps may move either progressively or episodically. A number of accounts document the sudden formation or reactivation of sackung scarps during large historical earthquakes. These data supports the idea that the geological record of surface ruptures associated with sackung could be used as secondary paleoseismic evidence.

8.9 Solifluction

Solifluction is a broad term used to describe a slow and shallow movement of soil in slopes. Some authors use creep as an equivalent term. Solifluction is related to two main deformation mechanisms within the soil. Cyclic heaving of particles normal to the slope and vertical settlement results in a net progressive downward displacement. This particle-scale motion may be related to freezing and thawing cycles, wetting and drying or swelling and shrinkage of clays. Additionally, the soil cover may slowly move downslope as a plug and undergo some internal deformation. This process is favoured by significant water content, high proportion fine-grained material and steep gradients. It is called gelifluction when the movement affects the active layer over the permafrost table. Solifluction rates are generally of the order of centimetres per year. Evidence of this sluggish but pervasive slope movement includes solifluction lobes, solifluction sheets and terracettes, as well as bent tree trunks (Figs. 8.9A, 8.9B and 8.9C).

Fig. 8.2A Section of a rotational landslide in an artificial exposure of a horizontally lying sequence composed of shales with interbedded limestone and gypsum. Note the inward dip of the mass displaced along a listric failure plane, Ebro Valley, NE Spain (Published with permission of © Francisco Gutiérrez 2015. All Rights Reserved)

Fig. 8.2B Rotational slide developed on the steep slope of a glacial valley consisting of limestones underlain by slates. Note the conspicuous headscarp and bench in the upper part of the landslide and the hummocky topography with closed depressions and limited drainage network in the lower part (Ampriu Valley, Cerler, Spanish Pyrenees) (Published with permission of © Francisco Gutiérrez 2015. All Rights Reserved)

Fig. 8.2C Active rotational landslide with a steep, arcuate headscarp and a backtilted upper surface. The landslide was induced by a large artificial excavation in the toe of the slope in Ojos Negros Mine, Iberian Chain, NE Spain (Published with permission of © Francisco Gutiérrez 2015. All Rights Reserved)

Fig. 8.3A Cinque Torri, Italian Dolomites, is a lateral spread developed on brittle dolomites underlain by soft marls. Some blocks are largely disrupted and affected by significant translation, subsidence and rotation (Published with permission of © Mauro Soldati 2015. All Rights Reserved)

Fig. 8.3B The grabens of Canyonlands, Utah, is an arcuate system of horst and grabens covering around 200 km^2 related to the lateral spread of a 460 m thick brittle sequence towards the Cataract Canyon of the Colorado River. The rock spread, with an estimated volume of around 60 km^3, is caused by the lateral flow of an underlying salt formation towards the unloaded canyon. This gravitational deformation has largely disrupted the pre-existing drainage (Image from © Bing Map)

Fig. 8.3C Wide-open fissures, grabens and sinkholes related to lateral spreading of a limestone unit underlain by slates, Benasque Valley, Spanish Pyrenees (Published with permission of © Francisco Gutiérrez 2015. All Rights Reserved)

Fig. 8.4A Debris flow track with lateral levees derived from a small debris slide, Loveland Pass, Southern Rocky Mountains of Colorado (Published with permission of © Francisco Gutiérrez 2015. All Rights Reserved)

Fig. 8.4B Multiple debris flows on forested debris cones in the Italian Dolomites (Published with permission of © Mauro Soldati 2015. All Rights Reserved)

Fig. 8.4C Fresh lobe-shaped debris flow accumulated on a debris cone at the fault-controlled western margin of the Dead Sea basin, Israel (Published with permission of © Francisco Gutiérrez 2015. All Rights Reserved)

Fig. 8.4D Small debris flow deposited on a low-gradient gully at the eastern margin of the Huerva River valley, NE Spain. Note the central depressed area generated by post-depositional water loss (Published with permission of © Francisco Gutiérrez 2015. All Rights Reserved)

Fig. 8.4E Slide in the Casitas Volcano, Nicaragua, transformed into a debris flow that incorporated a significant amount of sediment and water along its path. This lahar triggered on October 30, 1998 by the Hurricane Mitch, killed more than 2000 people in the towns of El Porvenir and Rolando Rodríguez (Published with permission of © A. Gómez-Sal 2015. All Rights Reserved)

Fig. 8.4F Earth flow developed on highly fractured Paleozoic slates in the Tena glacial valley, Spanish Pyrenees. Note the lateral levees and the lobate accumulation zone (Published with permission of © Francisco Gutiérrez 2015. All Rights Reserved)

Fig. 8.5A "Nefertiti rock" in the Desolation Canyon of the Green River, Utah, is an incipient topple developed on a well-bedded and subhorizontally lying calcareous formation (Published with permission of © Francisco Gutiérrez 2015. All Rights Reserved)

Fig. 8.5B Multiple rock topples in a gypsum escarpment in the Jalón River valley close to Calatayud City, Iberian Chain, NE Spain (Published with permission of © Francisco Gutiérrez 2015. All Rights Reserved)

Fig. 8.5C Soil topple on a cliff carved in an Holocene valley fill consisting of unconsolidated fine-grained deposits, Bardenas, NE Spain (Published with permission of © Francisco Gutiérrez 2015. All Rights Reserved)

Fig. 8.6A Image showing the scar and one of the blocks of a rock fall occurred on July 20, 2006, in Piz Somplunf, Badia Valley, Italian Dolomites. The 40,000 m^3 rock mass, detached from a cliff in a dolomitic massif, descended 800 m bouncing on rock slopes and a debris cone, progressively breaking up into smaller blocks. The impacts of the blocks on the loose cone produced conspicuous craters (Published with permission of © Francisco Gutiérrez 2015. All Rights Reserved)

Fig. 8.6B Talus slopes and long runout, large-sized blocks in the valley floor at Punta de Vacas, Mendoza Province, Argentinean Andes (Published with permission of © Mateo Gutiérrez 2015. All Rights Reserved)

Fig. 8.6C Rock falls derived from undermined limestone beds in a butte, Ksar of Ait Ben Haddou, Morocco (Published with permission of © Francisco Gutiérrez 2015. All Rights Reserved)

Fig. 8.6D Fallen sandstone blocks in a slope undermined by tafoni weathering, Ayers Rock (Uluru), central Australia (Published with permission of © Francisco Gutiérrez 2015. All Rights Reserved)

Fig. 8.7A Rock avalanche accumulated on a glacier in Mount Cook National Park, New Zealand (Image from © Bing Maps)

Fig. 8.7B Rock avalanche triggered by the M 7.9 Denali Fault, Alaska, earthquake of November 2002, and accumulated on the Black Rapids Glacier. The Denali Fault is buried by the glacier at this location. A large block failure from a granite slope transformed into a rock avalanche, crossed the ca. 2.5 km wide glacier including a 50 m high medial moraine and then turned and flowed downvalley (Published with permission of © Randall W. Jibson 2015. All Rights Reserved)

Fig. 8.8A Multiple uphill-facing scarps and associated troughs in the slope of the Vallibierna glacial valley, Spanish Pyrenees (Published with permission of © Enrique Acosta 2015. All Rights Reserved)

Fig. 8.8B Uphill-facing scarp and linear depression on its upslope side, Estós glacier valley, Spanish Pyrenees (Published with permission of © Francisco Gutiérrez 2015. All Rights Reserved)

Fig. 8.8C Ridge crest affected by gravitational spreading showing ridge-top depressions (grabens) bounded by uphill-facing scarps, Bear Mountain, Southern Rocky Mountains of Colorado (Published with permission of © Francisco Gutiérrez 2015. All Rights Reserved)

Fig. 8.9A Solifluction lobes on the southern slopes of the Ebba Valley, central Spitsbergen (Published with permission of © Zbigniew Zwolinski 2015. All Rights Reserved)

Fig. 8.9B Terracettes developed on slopes with a clay-rich regolith related to the weathering of highly fissile Paleozoic slates, Sierra Negra, Benasque Valley, Spanish Pyrenees (Published with permission of © Francisco Gutiérrez 2015. All Rights Reserved)

Fig. 8.9C Bent tree trunks that progressively restore their vertical orientation as they lean by solifluction (Published with permission of © Francisco Gutiérrez 2015. All Rights Reserved)

Fluvial Landforms

9

Fluvial landforms are those generated by running water, mainly rivers. The term fluvial derives from the Latin word *fluvius* that means river. Fluvial landforms cover an enormous range of dimensions, from small features like rills to major continental-scale morpho-hydrological units like large rivers and their drainage basins. The River Nile is 6,650 km long and the drainage basin of the Amazon River covers 7,050,000 km^2, an area almost as big as Australia. Rivers and streams drain most of the continental surface and occur in most environments, with the exception of some hyperarid regions, including vast sand seas, permanently frozen regions and karst terrains. Rivers flowing to the oceans drain about 68 % of the Earth's land surface. The drainage network plays a fundamental role in the continuous transfer of water and sediment from upland to lowland areas and from the continents to the oceans. Rivers transport approximately 75,000 million tonnes of material each year, of which 20,000 million tonnes reaches the sea, around 80 % in solid form and 20 % in solution. Fluvial systems are frequently the main agents involved in landscape evolution and exert a prime influence on other interrelated geomorphic systems like hillslopes, alluvial fans, deltas or beaches.

The drainage basin, also called catchment or watershed, is regarded by some authors as the fundamental geomorphic unit. This is the portion of land drained by a river, whose boundaries correspond to the drainage divide. The water that reaches the soil surface (e.g. rainfall, snowmelt) may either infiltrate or remain at the surface and run off downslope as overland flow. The infiltrated water may move through the soil cover roughly parallel to the ground surface as throughflow, or percolate downwards through the vadose or non-saturated zone until it reaches the phreatic or saturated zone, contributing to aquifer recharge and groundwater flow. Runoff occurs when the rainfall intensity exceeds the infiltration capacity of the soil and when the soil is totally saturated and cannot absorb any more water. Unconfined sheetwash in the slopes tends to concentrate in branching networks of channels, progressively larger as the contributing area increases (rills, gullies, streams, rivers). Perennial rivers hold water flow over most of the year, whereas ephemeral channels only show sporadic flow events. Most rivers flow into the oceans, lakes or inland seas. Some channels, especially in drylands and karst terrains, may dry up completely or disappear due to evaporation and/or infiltration. Channels may also grade into alluvial fans in piedmonts and tributary junctions.

Rivers show a great morphological diversity and may comprise one or multiple channels. Two main channel types with contrasting behaviour are differentiated depending on their substrate. Bedrock channels are cut directly into bedrock. Consequently they are characterised by a rather stable geometry. In contrast, alluvial channels that are formed in unconsolidated alluvium experience continuous changes by erosion and deposition processes. The morphological adjustments in these channels are controlled by a complex suite of interrelated factors like channel gradient, flow regime, sediment supply and the type sediment load (fine-grained suspended load versus coarse bedload). The behaviour of channels depends on the balance between the stream power and the power required to transport the sediment supplied to the stream (threshold of critical power). If their actual power exceeds such threshold, stream erosion prevails. Conversely aggradation dominates when sediment supply overwhelms the stream power. The main morphological changes in alluvial channels typically occur episodically during major flood events that interrupt relative long periods of limited change.

Rivers are the geomorphic systems that are of the greatest interest for the society. They are used for multiple purposes including water supply, transportation, power generation, waste disposal and recreation. They are also responsible for the natural hazard that causes the largest economic losses and great number of fatalities. Floodplains with their fertile soils and flat topography attract growing numbers of people. Many are protected by dikes designed to control ordinary floods, but these can be overtopped or fail by severe events putting large numbers of lives at risk. Human activity, notably through dam construction and land-use changes, contributes to significantly alter the regime and dynamics of rivers. These changes impact indirectly other interrelated

F. Gutiérrez, M. Gutiérrez, *Landforms of the Earth*, DOI 10.1007/978-3-319-26947-4_9

geomorphic systems, for example, the downstream deltas and beaches.

9.1 Bedrock Channel

These are channels cut into bedrock that lack a continuous cover of alluvial deposits. They develop where the transport capacity of the stream exceeds sediment supply, and consequently there is an energy surplus that can be spent in bedrock incision. Their existence may be determined by various factors such as high gradient, limited sediment supply and/or resistant bedrock. Bedrock channels are commonly narrow and lack a floodplain (e.g. canyons, gorges, ravines). In these constricted channels, flood flows experience dramatic increases in water depth and velocity and have a huge capacity to erode and transport boulders. Bedrock channels may be straight or display a sinuous path related to significant lateral migration during entrenchment, forming incised bedrock meanders (Figs. 9.1A, 9.1B and 9.1C). The main erosion processes in these channels, in which the water flow is in direct contact with bedrock, include block quarrying (plucking of rock blocks), abrasion by particles transported by rapid flows and dissolution in soluble rocks. Weathering processes may also contribute to rock wearing and the preparation of material for erosion. The longitudinal profile of bedrock channels may display pools and steps, frequently controlled by the presence of rocks with contrasting resistance. The channel bed may display cylindrical pits scoured in bedrock called potholes (Fig. 9.1D). They are produced by enhanced abrasion caused by coarse particles that swirl within vortices. Furrows and walls with sharp rounded concavities are also common. The banks of channels carved in soluble rocks may also display subhorizontal notches and scallops (Fig. 9.1E).

9.2 Knickpoint, Waterfall

Knickpoints are stream sections with substantially steeper gradient than in the upstream and downstream reaches. They are also referred to as headcuts, especially in alluvial channels, gullies and rills. Waterfalls are associated with subvertical drops in the longitudinal profile of the stream. Some authors contend that in waterfalls, in contrast with cascades and cataracts, the flow is dominated by free fall (Figs. 9.2A and 9.2B). The development of knickpoints and the associated waterfalls may be related to multiple processes like (1) differential incision controlled by rocks with contrasting resistance, (2) tectonic or isostatic uplift (e.g. hanging valleys on rock coasts), (3) dip-slip faulting, (4) glacial overdeepening, (5) upstream propagation of incision waves related to base-level falls and (6) construction of steps by tufa deposition. Erosional knickpoints retreat upstream. They may have a long persistence time if they evolve by parallel retreat or be ephemeral features when erosion processes rapidly wear away the crest of the fall. Several processes are commonly involved in the retreat of knickpoints. The falling water tends to excavate a plunge pool. Headward expansion of the pool by swirling flow undermines the rock face. The accelerating water flowing over the lip of the fall can detach blocks (by plucking) that may be loosened by unloading joints. Seepage and sapping erosion may also contribute to accelerate the recession of the rock face. When the undercut is separated from the waterfall, weathering expedited by abundant spray water may rapidly deteriorate the mechanical properties of the rock.

9.3 Alluvial Channel

Alluvial channels are essentially cut into unconsolidated sediments deposited by the fluvial system itself (alluvium). These channels are commonly flanked by nearly horizontal floodplains that become inundated when the flow exceeds the channel capacity (bankfull discharge). In contrast with bedrock channels, alluvial channels are highly dynamic systems that experience frequent morphological changes through erosion and deposition, especially during flood events. Long reaches of these channels, bounded by relatively soft banks, may abruptly modify their path through avulsion processes. The channel pattern depends on multiple factors, chiefly valley gradient, flood regime, sediment flux and type of sediment load (suspended load versus coarse bedload). Straight channels tend to be associated with low gradients or resistant banks that prevent lateral migration. Meandering channels, with a sinuosity higher than 1.3 (channel length–valley length ratio), tend to have cohesive banks, mixed sediment load and moderate width–depth ratios (Fig. 9.3A). The external concave bank is dominated by erosion, whereas deposition occurs in the inner convex bank, mainly by lateral accretion, resulting in the formation of point bars. The meander loops may be cut off, leading to the development of oxbow lakes in the abandoned channel sections (Fig. 9.3B). Braided channels are characterised by multiple bars, exposed at low flow, interlaced by a braided network of channels. These are typically bedload-dominated channels with high width–depth ratios commonly associated with relatively steep and high energy systems (Fig. 9.3C). Anabranching rivers consist of multiple channels separated by vegetated and relatively stable floodplain portions. The term anastomosing river is generally restricted for low-energy anabranching systems dominated by fine-grained or organic deposition (Figs. 9.3D and 9.3E).

Fig. 9.1A Canyon-like bedrock channel deeply incised into calcareous conglomerates, Cijevna River, Montenegro (Published with permission of © Francisco Gutiérrez 2015. All Rights Reserved)

Fig. 9.1B Tortuous incised meanders of the San Juan River at the Goosenecks State Park, Colorado Plateau, Utah. This bedrock channel has downcut with significant lateral migration into subhorizontally lying sedimentary formations in response to relatively slow regional uplift (Image from © Bing Maps)

Fig. 9.1C The Black Canyon of the Gunnison, Colorado, carved into resistant Precambrian metamorphic rocks. Note the tilted and planar unconformity on the left, exhumed after the erosion of a Mesozoic sedimentary cover (Published with permission of © Francisco Gutiérrez 2015. All Rights Reserved)

9.4 Ephemeral Channel

Ephemeral channels are those that remain dry most of the year and show sporadic short-lasting flow events. They are most common in arid and semiarid regions, where a significant part of the annual precipitation typically falls through storm events. They are designated with numerous terms, depending on the geographical region, such as arroyo and wash in the USA, wadi in northern Africa and Arabia, nullah in India, donga in South Africa or barranco and rambla in Spain. They are frequently alluvial channels that may be cut into older alluvial fills (valley fills). Flow events are commonly related to intense and/or high-magnitude rainfall events. In contrast with perennial streams, discharge may decrease significantly downstream and disappear despite the increase in catchment area, due to infiltration into the channel floor (an influent stream). These fluvial systems are typically affected by sudden and short duration flash floods, characterised by very high peak discharge and abundant sediment load. These catastrophic events may have a dramatic geomorphic effectiveness, producing drastic changes in the channel geometry and path by erosion and/or aggradation. Moreover, in some regions, flash floods in these fluvial systems, frequently perceived as harmless or inactive, are responsible for the largest number of flood-related fatalities. Ephemeral channels are highly sensitive to climate variability, active tectonics and human alterations in the landscape (e.g. vegetation removal), and they commonly have experienced multiple aggradation and incision phases during the Holocene (Figs. 9.4A and 9.4B).

9.5 Terrace

Fluvial terraces are former valley floors that become abandoned by river entrenchment. Terraces surfaces, called treads, are characterised by a planar geometry with a slight downstream inclination. The typically scarped valley-facing

Fig. 9.1D Potholes carved in granitoids and largely controlled by jointing in a reach with rapids of the Tietar River, La Iglesuela, Central System, Spain (Published with permission of © José Ortega 2015. All Rights Reserved)

Fig. 9.1E Five nested subhorizontal notches carved on folded salt of the Gachsaran Formation during a flood, Karun River, Ambal, Zagros Mountains, Iran (Published with permission of © Mohammad Hasan Nazari 2015. All Rights Reserved)

edge is known as the riser. Terrace surfaces may have a depositional or erosional nature. Fill terrace refers to relict surfaces underlain by relatively thick alluvium that records fluvial aggradation. Erosional terraces (cut terraces or strath terraces) may develop by lateral migration and valley-floor widening during phases of base-level stability. These surfaces may be cut into older alluvium (fill-cut terrace) or across bedrock (rock-cut terrace). Rock-cut or rock-strath terraces may be underlain by a thin veneer of channel deposits capping a planar bedrock erosional surface. The terrace surface and the associated fluvial deposits may be buried by younger sediments like colluvium, pediment deposits, loess, volcanic ash or overbank facies related to extraordinary floods. Fluvial valleys commonly display multiple terrace levels that record the long-term entrenchment history of the river, interrupted by phases of aggradation and/or lateral migration. Main controls on terrace development include water discharge, flood regime and sediment flux, all of which respond to climatic factors and, for recent times, anthropogenic activity. In addition, gradients may be affected by tectonics and other causes of base-level change. The distribution of the terraces provides information on their relative chronology, with the oldest terraces at the highest elevation with respect to the current channel. Dated terrace sequences may be used to infer the evolution of the landscape, including long-term incision rates. They also constitute excellent geomorphic markers for the identification and assessment of neotectonic deformation (Figs. 9.5A and 9.5B).

9.6 Alluvial Fan

Alluvial fans are depositional landforms with a semiconical geometry that develop in piedmonts and at tributary junctions, where the feeder channel loses confinement favouring rapid deposition. Alluvial fans are characterised by convex cross profile and relatively high slopes that radiate away from the apex (Fig. 9.6A). The proximal fan sector

Fig. 9.2A Lower Falls around 100 m high in the Grand Canyon of the Yellowstone River, Yellowstone National Park, Wyoming. This knickpoint is controlled by a ledge of resistant rhyolite (Published with permission of © Francisco Gutiérrez 2015. All Rights Reserved)

Fig. 9.2B The Gullfoss waterfalls in the canyon of the Hvita River, Southern Iceland. Here, the trace of this bedrock channel is controlled by extensional structures related to mid-ocean spreading (Published with permission of © Francisco Gutiérrez 2015. All Rights Reserved)

Fig. 9.3A The Aguas Tuertas meandering river developed in a sediment-filled glacial overdeepened basin, Guarrinza Valley, Spanish Pyrenees (Published with permission of © Francisco Gutiérrez 2015. All Rights Reserved)

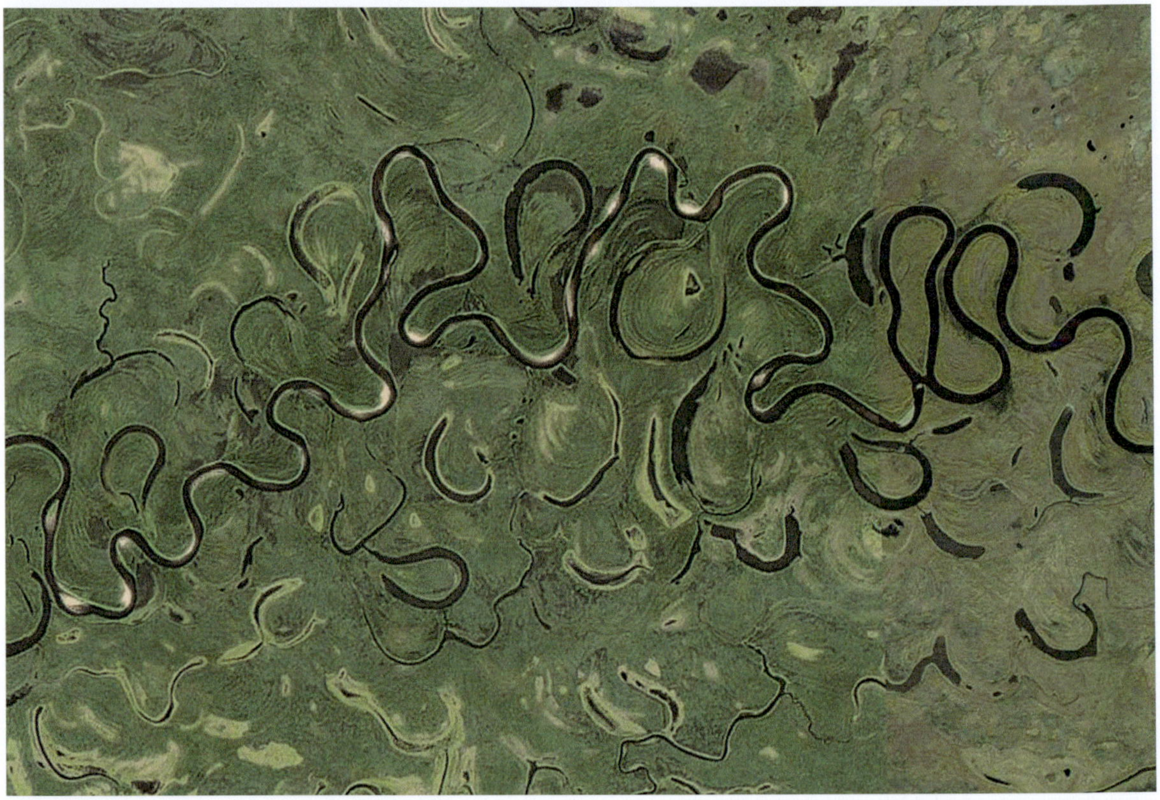

Fig. 9.3B Multiple abandoned meander loops in the floodplain of the Black River, Alaska (Image from © Bing Maps)

Fig. 9.3C The Rakaia River, New Zealand, in a section with braided pattern. This bedload-dominated channel, with a large width–depth ratio, shows multiple interlaced channels and intervening longitudinal gravel bars (Image captured from Google Earth)

Fig. 9.3D The Ob River, Russia, displays multiple anabranching channels carved in a broad floodplain (Image from © Bing Maps)

Fig. 9.3E The Ebro River valley during the 2015 flood, with a return period of around 10 years. The image illustrates that building in floodplains is not a wise concept. Zaragoza city in the background (Published with permission of © Francisco Gutiérrez 2015. All Rights Reserved)

may be incised by an entrenched channel whose downstream limit is called the intersection point. Fanhead trenching promotes distal aggradation and the progradation of the fan. The main processes involved in the construction of alluvial fans are sheetfloods and debris flows (Figs. 9.6B and 9.6C). These depositional processes are typically infrequent catastrophic events related to heavy rainfall events. The sediment-laden water flows become unconfined when they reach the fan apex or the intersection point and expand sheetflooding the active lobe of the fan. Flow expansion entails an abrupt decrease in the depth, velocity and competence of the flow, inducing rapid deposition. Sheetflood deposits typically consist of sheetlike beds of angular, poorly sorted gravelly deposits with a syndepositional dip roughly parallel to the fan surface. In contrast, debris flows mainly produce paired levees and radially elongated lobes usually consisting of a coarsening-upward, matrix-rich gravelly deposit with chaotic fabrics. The deltas whose subaerial component is an alluvial fan are called fan deltas (Fig. 9.6D).

9.7 Mantled Pediment

Mantled pediments, called accumulation glacis by the French geomorphologists, are piedmont surfaces gently inclined towards the base level (e.g. river, lake), underlain by an alluvial cover typically only a few metres thick and generally developed on relatively soft bedrock. They typically display a planar geometry and a concave longitudinal profile, with the slope progressively increasing towards the proximal sector. The cover is typically a subangular, poorly sorted gravel deposit dominated by sheetlike beds. The lithological spectrum, in contrast with that of the fluvial terraces, is constrained by the rocks exposed in the restricted source area. The formation of mantled pediments seems to record relatively long periods of base-level stability, during which the marginal fluvial systems, once they reach an equilibrium profile, expand laterally and experience poorly confined sheetflooding in response to heavy storms. Mantled pediments are particularly frequent in semiarid environments.

Fig. 9.4A Ephemeral channel (wadi) west of Hail, northern Saudi Arabia. Water flow rapidly infiltrates in these valleys, commonly underlain by pervious alluvium perched above the water table (influent stream) (Published with permission of © Francisco Gutiérrez 2015. All Rights Reserved)

They may form stepped sequences and in some cases can be correlated with terraces. The development of duricrusts may substantially enhance the preservation potential of mantled pediments, which eventually may form prominent surfaces through a process of relief inversion (Figs. 9.7A and 9.7B).

9.8 Rill

Rills are small erosional channels developed on slopes with a maximum width and depth of a few tens of centimetres. They typically form on bare or sparsely vegetated surfaces underlain by low-permeability soils and affected by frequent intense rainfall and overland flow. During a rainfall event, the raindrops that impact on bare unprotected surfaces may have sufficient momentum to detach particles and redistribute them radially. This process of rainsplash erosion involves a net downslope movement of particles. When the rainfall rate exceeds the infiltration capacity of the soil, the water ponds and runs off downslope. In the upper part of the slope, overland flow is essentially an unconfined sheetwash with very limited erosion capacity. With increasing distance from the divide, as the contributing area augments, the flow depth and the shear stress exerted on the surface rise to reach threshold values for the erosion of loose weathered material. Erosion may create preferential flow paths where the overland flow concentrates producing shallow rills. The rills may enlarge through a self-accelerating process, whereby flow concentration promotes higher erosion, which in turn leads to higher flow and erosion. Erosion in the rills is caused by concentrated flow, whereas unconfined sheetwash and rainsplash are the dominant processes in the inter-rill areas. Downslope, as run-off increases, the erosion stress may be high enough to incise into the underlying material to cut gullies (Figs. 9.8A and 9.8B).

9.9 Gully

Gullies are small and deeply incised channels carved by overland flow episodes, generally related to severe rainfall events. They typically have steep sides and an abrupt upstream termination called the gully head. Badlands refers to erosional landscapes dissected by a dense network of

Fig. 9.4B Ephemeral stream (wadi) after a flash flood that occurred in 2014 in the Atlas Mountains, Morocco (Published with permission of © Francisco Gutiérrez 2015. All Rights Reserved)

Fig. 9.5A Stepped sequence of fluvial terraces in the Alfambra River valley, Iberian Chain, Spain. The terrace deposits consist of a relatively thin alluvial cover underlain by Tertiary claystones (red) and gypsum (white). Old terrace surfaces, once situated in the valley floor, form prominent buttes capped by indurated alluvium (relief inversion) (Published with permission of © Francisco Gutiérrez 2015. All Rights Reserved)

Fig. 9.5B Terraces in the Gállego River valley, Ebro Cenozoic Basin, NE Spain. Youngest levels mainly preserved in the inner side of meanders (Published with permission of © Francisco Gutiérrez 2015. All Rights Reserved)

Fig. 9.6A Large alluvial fan in the Gobi Desert, NW of Chijinpu, China (Image from © Bing Maps)

Fig. 9.6B Sheetflood-dominated alluvial fan in Death Valley National Park, California (Published with permission of © Francisco Gutiérrez 2015. All Rights Reserved)

Fig. 9.6C Debris flow-dominated alluvial fan in Death Valley National Park, California. Note the relative age of the lobes indicated by the degree of desert varnish development; the darker the older (Published with permission of © Francisco Gutiérrez 2015. All Rights Reserved)

Fig. 9.6D Fan delta at the mouth of Wadi Mujib in the Jordanian side of the Dead Sea. In the 1950s, the lower part of the canyon was permanently flooded, forming an inlet of the Dead Sea. The fan delta has developed in response of the rapid anthropogenic lowering of the Dead Sea level, at a rate of around 1 m/year during the twenty-first century (Published with permission of © Damien Closson 2015. All Rights Reserved)

Fig. 9.7A Perched mantled pediment cut across folded marls and mantled by a thin alluvial cover, Canal de Berdún, Spanish Pyrenees (Published with permission of © Francisco Gutiérrez 2015. All Rights Reserved)

Fig. 9.7B Lateral view of two mantled pediment levels in the Draa Valley, southern Morocco (Published with permission of © Francisco Gutiérrez 2015. All Rights Reserved)

Fig. 9.8A Rills carved on Tertiary claystones and micropediment at the foot of the slope, Bardenas Reales, northern Spain (Published with permission of © Francisco Gutiérrez 2015. All Rights Reserved)

Fig. 9.8B Crop field affected by rilling, favoured by ploughing parallel to the slope gradient, Bardenas Reales, northern Spain (Published with permission of © Francisco Gutiérrez 2015. All Rights Reserved)

gullies (Fig. 9.9A). Gullying typically occurs, but not exclusively, in semiarid regions with limited vegetation cover and a relatively soft substrate (e.g. unconsolidated surficial formations, residual soils, argillaceous bedrock, fine-grained volcanoclastic rocks). According to their geomorphic position, two main types may be differentiated: hillslope gullies and valley-floor gullies, commonly with lower gradients than the former. The formation of gullies and badland landscapes may be initiated by a number of factors, including reduction in the vegetation cover by natural or anthropogenic causes, base-level lowering, severe rainfall events or landsliding. In addition to erosion by concentrated run-off in the channels, other processes may significantly contribute to the development and enlargement of gullies. Weathering processes, including wetting-drying and freezing-thawing cycles, or carbonate dissolution in marls, lead to the development of a weak regolith preparing material for erosion during subsequent flow events. The weathered mantle may be affected by spalling and shallow slides that disintegrate and transform into mud flows, which may partially fill the gullies downslope (Fig. 9.9B). The contraction associated with desiccation, especially in sediments with swelling clays, produces substantial surface cracking that facilitates water infiltration and subsurface erosion (piping). New gullies result from the collapse of the pipes. The sides of the gullies may be also oversteepened by rapid downcutting and/or undermining, promoting the development of mass movements. Sapping erosion may also contribute to the headward propagation of the gully head. Differential erosion controlled by resistant beds or large particles may lead to the formation of pillars, also known as hoodoos or pedestals (Fig. 9.9C). Gullies are highly dynamic systems whose evolution is largely controlled by their connectivity with the basal channel and the condition of the base level. Effective hillslope–channel coupling allows the deposits derived from the gully system to be flushed away, favouring further hillslope incision. A falling base level involves an overall gradient increase promoting downcutting and headward expansion of the gully systems. Under a stable local base level, the slopes evolve by parallel retreat resulting in the progressive development of pediments at their foot.

Fig. 9.9A Badland landscape with hillslope and valley-floor gullies in Bardenas Reales, northern Spain (Published with permission of © Francisco Gutiérrez 2015. All Rights Reserved)

Fig. 9.9B Mud flows in gullies derived from shallow regolith slides, Bardenas Reales, northern Spain (Published with permission of © Francisco Gutiérrez 2015. All Rights Reserved)

Fig. 9.9C Gullies and hoodoos developed in softened Paleozoic rocks associated with a Quaternary fault, Los Abuelos de Tobed, Iberian Chain, NE Spain (Published with permission of © Francisco Gutiérrez 2015. All Rights Reserved)

9.10 Piping

Piping refers to subsurface water erosion in soils and unconsolidated deposits resulting in the development of shallow pipes. The conduits may reach several metres in diameter and form well-developed subsurface drainage networks that in some cases can be explored and mapped. They are frequently described as pseudokarst phenomena because of the associated internal drainage and collapse sinkholes. Pipes frequently form by focused erosion at seepage outlet points, where water pressure is high enough to mobilise fine particles. In this case, the seepage outlet and the conduits propagate backwards. Pipe development may also start at cracks or other discontinuities at the surface and expand towards the outlet of the seepage flow. The inlet points may correspond to desiccation cracks, unloading cracks, landslide-related fissures, tree roots or animal burrows. Once the conduits reach a critical size for which the flow changes from laminar to turbulent, erosional enlargement may be very rapid. The main factors that favour piping development include (1) weak and porous fine-grained sediments; (2) an impeding layer that deflects laterally seepage flow; (3) sufficient water supply; (4) steep hydraulic gradients, typically associated with slopes and scarps; and (5) development of surface cracks. In argillaceous sediments, surface cracking and piping are favoured by the presence of swelling and dispersive clays. Swelling clays (e.g. montmorillonite) experience significant volume changes upon wetting and drying. Dispersive clays are highly susceptible to deflocculation, involving the disaggregation of individual particles that can be easily washed away. The pipes may evolve into chains of sinkholes by roof collapse, which ultimately lead to the development of gullies that may be locally interrupted by soil arches (Figs. 9.10A and 9.10B). Widespread piping and collapse may produce peculiar landscapes riddled by holes and intervening pinnacles (Fig. 9.10C). Piping may also play a significant role in landslide development. It contributes to reduce the shear strength of potential failure surfaces and may create water overpressure conditions when the input of water exceeds the discharge capacity of the conduit network. Piping, also known as tunnel erosion, is one of the main causes of failures in earth dams, embankments and flood-control dykes.

Fig. 9.10A Collapse sinkhole related to piping at the edge of a valley-floor gully, Bardenas Reales, northern Spain (Published with permission of © Francisco Gutiérrez 2015. All Rights Reserved)

Fig. 9.10B Section of a perched and abandoned pipe that has guided the development of a gully, Bardenas Reales, northern Spain (Published with permission of © Francisco Gutiérrez 2015. All Rights Reserved)

Fig. 9.10C Pinnacled topography developed in a valley fill related a dense network of subsurface pipes and their collapse, Aguarales de Valpalmas, NE Spain (Published with permission of © Francisco Gutiérrez 2015. All Rights Reserved)

9.11 Armoured Mud Ball

Armoured mud balls, also called mud balls, are subspherical particles consisting of a core of fine-grained cohesive material, coated by clasts. They typically have a centimetre-scale diameter, but may reach more than 10 cm. They are mainly found associated with badlands and ephemeral streams. The balls commonly originate as clayey chunks incorporated into a stream by bank erosion (rip-up clast). As the clay clast rolls along the channel as bedload during a flood, its outer part is softened upon wetting and enlarges by the accretion of additional mud with embedded armouring clasts. The development of mud balls is favoured by seasonal semiarid climate. Wetting and drying cycles promote the development of an erodible regolith with desiccation cracks for the production of rip-up clasts. Another important factor is slope–channel connectivity to allow the incorporation of clay chunks into the stream during flood events (Fig. 9.11).

Fig. 9.11 Recently formed bar of armoured mud balls, Bardenas Reales, NE Spain (Published with permission of © Francisco Gutiérrez 2015. All Rights Reserved)

Coastal Landforms

10

The coast is the zone where the land and the sea interact. Here, waves, tides and nearshore currents, together with fluvial processes, wind action and reef development, generate depositional and erosional landforms. The coastline is the land margin defined by the normal high spring tides. It is estimated that the total length of the coastline in the world is more than a million kilometres long. The coasts can be divided into different zones according to hydrodynamic criteria with morphogenetic implications. The boundaries of these zones change through time with variations in the wave and tidal regimes. The shore is the zone between low-tide level and the landward limit of wave action. It comprises the backshore and the foreshore, beyond and below high-tide level, respectively. The backshore may be sporadically inundated by extraordinary high tides or large storm waves. The nearshore is the shallow water zone situated above the low-tide line and significantly affected by wave action, including the surf zone and the swash zone. The offshore zone extends below the low-tide level.

The position of the coast has experienced significant changes during the Quaternary, especially in low-lying littoral zones and in tectonically active regions. The coastline may advance when (1) the accumulation of deposits exceeds the erosion rate (progradation); (2) tectonic uplift causes the emergence of new land; and (3) there is a sea-level fall, mostly related to glacial periods (glacio-eustasy), during which a significant proportion of the oceanic water is transferred to the continents in the form of glaciers. During the Last Glacial Maximum, at ca. 21,000 years ago, when glaciers used to occupy around 25 % of the Earth's continental area, the sea level was approximately 125 m lower than at the present time. The opposite processes, net erosion, tectonic subsidence and sea-level rise result in the retreat of the coastline.

Waves, tides and nearshore currents are the main morphogenetic processes acting on coasts. Waves are undulations in the water surface related to the transfer of energy form the wind to the seawater. The waves approaching the shallow nearshore slow down and increase in height, and their front steepens and eventually breaks in the surf zone. The oscillatory movement in the offshore zone changes into a translational forward (swash) and backward (backwash) movement in the swash zone. Breaking waves, depending on their characteristics, may cause a net shoreward or a net seaward movement of sediment. Waves that approach the coast obliquely experience refraction and generate long-shore currents. These currents cause a net longitudinal transport of sediment and may generate spits associated with sharp changes in the orientation of the coast. Tides are the rise and fall of the sea level caused by the combined effects of gravitational forces caused by the Moon and the Sun, and the rotation of the Earth. In most places they show a semidiurnal periodicity, with two cycles each day. The higher than normal high tides that occur when the Moon and the Sun are aligned are called spring tides. Three tidal ranges are differentiated: microtidal (<2 m), mesotidal (2–4 m) and macrotidal (>4 m). A large tidal range in low-gradient coasts sets aside wave action and favours the development of estuaries and tidal flats with high-energy tidal currents.

Humans can significantly alter coastal processes and can be also severely affected by them. More than half of the world's population resides in coastal regions, and many people visit these zones mainly for recreation. A good understanding of coastal processes and their geomorphic setting is an essential basis for proper sustainable coastal management. Some of the main human impacts on coastal environments include the following: (1) Changes in the sediment load in rivers, mainly reductions related to dam construction. The latter may lead to rapid retreat in deltas and beaches. (2) Construction of protective structures (e.g. sea walls, groynes) that modify sediment dynamics, with potential adverse effects even in distant areas. (3) Significant changes in the configuration of the coast by land reclamation. For instance, nearly 17 % of the Netherlands are lands reclaimed for the sea. (4) An ongoing sea-level rise induced by the anthropogenic global warming. Some of the expected effects of the rising sea levels include (1) submergence of low-lying coastal areas; (2) enhanced cliff-base erosion resulting in

F. Gutiérrez, M. Gutiérrez, *Landforms of the Earth*, DOI 10.1007/978-3-319-26947-4_10

more frequent landslide activity and more rapid cliff retreat; (3) higher erosion in beaches; (4) breaching and erosion of barriers and spits enclosing lagoons, which eventually may be transformed into open inlets or embayments; and (5) renewed vertical accretion in reef platforms to keep pace with the sea-level rise.

10.1 Coastal Cliff

Coastal cliffs are steep slopes, frequently subvertical, mostly developed on resistant rocks. In most cases their formation is related to basal wave erosion and rapid retreat. Some coastal cliffs are largely related to other processes like dip-slip faulting or previous inland erosion (e.g. steep walls of glacial troughs invaded by the sea). Basal erosion by wave attack creates an oversteepened slope that may display a linear horizontal notch at its foot. Notches indicate the high- or mid-tide level and are particularly well developed in limestone and in coasts with a restricted tidal range, where undermining is concentrated within a restricted elevation range. The overlying undermined material may collapse through various types of mass movements, and the landslide deposit accumulated at the cliff foot may be rapidly removed by marine erosion. The profile of the cliffs is largely related to the relative effectiveness of marine and subaerial processes, which dominate in the lower and upper parts of the slopes, respectively. Steep cliffs develop where marine processes dominate (e.g. waves, bioerosion, wetting and drying cycles, salt weathering). Cliffs have lower gradients where subaerial process prevails (e.g. run-off erosion, mass movements), like in areas with weak waves. Differential cliff retreat along the coast, commonly controlled by lithological changes and tectonic structures, may result in the development of irregular coastlines with headlands and bays associated with more resistant and more erodible material, respectively. Differential cliff retreat and the dissection of headlands may isolate monolith-like islets known as stacks. Narrow headlands and slablike stacks may be pierced by preferential cliff-base erosion leading to the development of natural arches. Arches in karst rocks may be also related to pre-existing dissolution caves truncated by erosion (Figs. 10.1A, 10.1B, and 10.1C).

10.2 Shore Platform

Shore platforms, formerly designated by the genetic terms wave-cut platforms or abrasion platforms, are erosional rock surfaces resulting from the retreat of coastal cliffs. These are time-transgressive straths, progressively older towards the seaward edge of the platform. Two main morphological types are differentiated: subhorizontal platforms and sloping platforms. Subhorizontal platforms commonly terminate at their seaward edge with a marked cliff that drops into the sea (low-tide cliff) and tend to develop in regions with low tidal range, which concentrates erosion within a restricted elevation range. Sloping platforms typically have a gradient between 1° and 5° and are more frequent on macrotidal coasts. A wide range of subaerial and marine processes may be involved in the formation of shore platforms, notably mechanical wave erosion, weathering (e.g. salt weathering, wetting and drying) and bioerosional activity. Wave erosion includes several mechanisms such as abrasion (friction caused by moving particles), impact forces, hydrostatic pressure and compression of air in joint spaces. Some shore platforms may correspond to inherited rock-cut benches developed during previous interglacial periods, when the position of the sea level was similar to that of the present time, or to raised platforms formed at lower elevations (Figs. 10.2A and 10.2B).

10.3 Marine Terrace

Marine terraces are relatively flat relict surfaces of marine origin situated above the current sea level. These perched benches may be horizontal or gently inclined and may form steeped chronosequences. Marine terraces may correspond to rock-cut surfaces generated by marine erosion, or depositional surfaces built by the accumulation of sediments in a former shore environment, including raised beaches. Some authors restrict the term to the latter description. In intertropical regions, marine terraces may also result from construction of coral reefs and the accumulation of reef material. In some regions (e.g. northern Spain), planar and gently sloping coastal terraces, which may have a mixed marine and subaerial origin, are called rasas, a Spanish word meaning flat. Marine terraces frequently record relatively warm interglacial periods characterised by restricted glacial ice masses and relatively high sea level. Quaternary terraces are restricted to elevations only a few metres above present sea level, unless they have been uplifted. High terraces on stable regions (e.g. passive plate margins) are much older and formed when sea level was tens of metres above today's level. The elevated position of marine terraces may be related to different mechanisms: (1) tectonic uplift and/or coseismic uplift, (2) glacioisostatic rebound, (3) eustatic sea-level fall and (4) depositional regression. In the two former situations, the geomorphic surface was developed at a lower elevation. Marine terraces are proxies of paleo-sea level. Numerically dated marine terraces and terrace sequences may be used to characterise neotectonic deformation, calculate uplift rates and reconstruct past sea-level changes (Figs. 10.3A, 10.3B, and 10.3C).

Fig. 10.1A Natural arch developed in a narrow headland consisting of subhorizontal Oligocene-age Lower Coralline Limestone in Gozo Island, Malta. Note the curved tension crack on the left flank of the arch and the basal notch in the plunging cliff (Published with permission of © Francisco Gutiérrez 2015. All Rights Reserved)

10.4 Beach

A beach is an accumulation of unconsolidated detrital deposits, ranging in size from fine sand to cobble gravel or even boulder, mainly formed by waves along the shore. Beaches occur along about 30 % of the world's coastline. They may show a wide variety of geometries and dimensions, from long straight beaches with broad surf zones to curved pocket beaches in bays and coves flanked by headlands that refract the waves (Figs. 10.4A and 10.4B). The beach profile comprises three zones with distinctive processes and geomorphic features: (1) The shoaling zone, between the wave base (offshore limit of wave action) and the breakpoint, where waves increase their height and entrain and transport sediment towards the shore. (2) The surf zone that extends from the breakpoint to the shoreline, where the waves break and turbulence increases. This zone may display one or several ridge and runnel complexes, consisting of longitudinal bars and associated troughs on the landward side (Fig. 10.4A). (3) The swash zone, affected by wave swash and backswash that may culminate in one or several berms (Fig. 10.4C).

This portion of the beach face may display evenly spaced cusps consisting of crescent-shaped embayments and horns with coarser deposits (Fig. 10.4D). The upper edge of the beach face may be defined by a scarp generated by erosional storm events. The backshore is the upper and commonly dry part of the beach, located beyond the normal swash limit. It is typically a subhorizontal or gently landward sloping surface. The backshore is affected by sporadic storm waves that may produce beach ridges and is frequently an area of eolian deposition. The main sources of beach sediment include fluvial supply, erosion from cliffs and rocky shores and the sea floor. These deposits may be transported and redistributed along the coast by long-shore currents. Beaches are highly dynamic and sensitive morpho-sedimentary systems that may experience abrupt changes in response to storms, tsunamis and human alterations.

Fig. 10.1B Dwejra Bay in Gozo Island, Malta, is a subcircular bay related to differential erosion of the softer sediment (Globigerina Limestone) downdropped within a paleocollapse structure controlled by subvertical ring faults. The cliffs on the Lower Coralline Limestone essentially correspond to an exhumed gravitational annular fault plane. Note the stack in the seaward edge of the bay (Published with permission of © Francisco Gutiérrez 2015. All Rights Reserved)

10.5 Spit and Barrier

Spits are narrow and elongated beach accumulations attached to the land at one end that mainly grow by long-shore sediment deposition. Their formation typically starts at re-entrants and sharp trend changes in the coastline, where waves arriving obliquely to the shore undergo refraction and energy dissipation, accompanied by sediment accumulation (Fig. 10.5A). The spit is attached to the coast in its proximal edge, where it is initially nucleated, and its extension mainly occurs by sediment addition at its terminus. Spits frequently end with one or several landward hooks or recurves, related to waves that arrive at different directions and to wave refraction around the end (Fig. 10.5B). These features are enhanced during periods of reduced long-shore sediment supply. The term barrier refers to depositional features that tend to bridge and connect the opposing ends of a re-entrant or bay (Fig. 10.5C). They may have evolved from spits, although some barriers are thought to be mainly related to onshore sediment supply, rather than long-shore drift. Spits and barriers may be capped by eolian dunes. Barriers are frequently interrupted by breaches and thus compartmentalised into barrier islands. These water inlets allow permanent or intermittent water exchange between the sea and the back-barrier environments. Spits and barriers create a sheltered zone on the landward side dominated by fine-grained deposition under low-energy conditions. They may be breached by high-energy storm or tsunami events, generating wash-over fans in the landward side.

10.6 Lagoon

Lagoons are water bodies in basins partly or completely separated from the sea by narrow ridges that may correspond to spits, beach barriers or coral reefs (Figs. 10.6A and 10.6B). Water bodies that are fully enclosed and disconnected from the sea are also called coastal lakes. The lagoon barrier commonly displays entrances that allow the exchange of water with the sea. These entrances may be gaps between spits and the mainland or channels breached by storms, tsunamis or floods from fluvial systems. The persistence of these

Fig. 10.1C Limestone stack (tower karst) with basal notch in Halong Bay, Vietnam (Published with permission of © Javier Ventosa 2015. All Rights Reserved)

entrances largely depends on the balance between the amount of sediment accumulated by wave action and the amount flushed by tidal currents. The formation of lagoons is mainly related to spit growth or the building up of shoals or bars at the mouth of drainages or the distal edge of embayments. Lagoons are typically shallow and low-energy sedimentary environments protected from ocean waves and dominated by fine-grained deposition. Main sources of sediments include fluvial supply from the mainland; input of sediment from the sea, including the accumulation of wash-over fans during high-energy events; autochthonous organic sedimentation; and wind-blown sediment. Aggradation may lead to rapid reductions in depth and size in these commonly shallow environments. Shallow lagoons and coastal lakes with evaporite deposition in the hyperarid coasts of the Arabian Peninsula are known as sabkhas, from the Arabic sebkha (Figs. 10.6A, 10.6B, and 10.6C).

10.7 Tombolo

A tombolo is a ridge of sand or gravel (shingle) built by wave action that links an island or a stack with the mainland. Its formation may be related to the growth of spits by long-shore deposition. Tombolos may be also related to deposition in the sheltered lee zone between a nearshore island and the mainland, where refracted waves undergo energy dissipation. There are examples of multiple tombolos that connect two or more islands. A tombolo that is fully exposed only at low tide is called tie bar (Figs. 10.7A and 10.7B).

10.8 Estuary

Estuaries are semi-enclosed coastal areas commonly associated with the mouth of river valleys and affected by tidal oscillations, where there is an intermixing of saline water from the sea and freshwater from land (Fig. 10.8A). The denser saline water may intrude as a wedge beneath the freshwater. A great part of modern estuaries are related to the inundation of the lower reaches of fluvial valleys due to post-glacial sea-level rise, which reached its approximate present level around 6,000 years ago. During the last glacial period, when the sea level was more than 100 m lower than at the present time, many rivers incised deep valleys and their mouths were in more distal sectors. The formation of estuaries may be also related to the inundation of glacial troughs in fjords and tectonic subsidence. The morpho-sedimentary

Fig. 10.2A Shore platform cut across steeply dipping flysch sediments in Zumaia, northern Spain (Published with permission of © Francisco Gutiérrez 2015. All Rights Reserved)

features of estuaries largely depend on the relative role played by waves, tides and fluvial processes, with a relatively higher influence in the mouth, medial and head sectors, respectively. The most distinctive processes of estuaries are periodic incoming (flow) and outcoming (ebb) currents controlled by tidal oscillations. These tidal ebbs and flows carve complex channel systems in low-relief mudflats and sandflats (tidal flats) that are exposed at low tide. Tide-dominated estuaries are characterised by dominant sediment transport towards the head and a decrease in sediment size in the same direction. Salt marshes typically develop at the head and in the more elevated sectors of the estuaries. They are characterised by fine and organic-rich deposits as well as halophilous vegetation (Figs. 10.8B, 10.8C and 10.8D). In tropical and subtropical areas, the intertidal estuarine zones are typically colonised by mangroves with dense systems of aerial roots that decelerate currents and prevent erosion during storm events (Figs. 10.8A, 10.8B, 10.8C and 10.8D).

Fig. 10.2B Shore platform carved in strata dipping seaward at Traeth Mawr Geopark, Wales (Published with permission of © Paula J. James 2015. All Rights Reserved)

10.9 Ria

Ria, a term derived from northwestern Spain, designates long and frequently branching coastal inlets generated by the partial submergence of deeply entrenched fluvial valleys. The valleys, originally produced by fluvial incision, were drowned during the postglacial sea-level rise (Holocene transgression). This base-level rise has promoted fluvial and marine sediment aggradation in these environments. Rias can be considered as a type of estuary, in as much there is freshwater and saltwater mixing, as well as tidal oscillations. Rias are narrow inlets typically bounded by steep slopes (Fig. 10.9).

10.10 Delta

Deltas are coastal landforms formed at the mouth of rivers by deposition of sediments mostly supplied by the fluvial system. Their development and evolution mainly depend on the balance between the constructional role played by riverine sediment flux and the destructional effect of marine processes, mainly waves. In deltas, the sediment-laden fluvial flows become unconfined, spread radially and experience an abrupt competence loss leading to rapid deposition. Delta sediments typically display an overall increase in grain size towards the proximal sector. The archetypal delta shows a relatively flat delta plain traversed by multiple distributary channels. The upper delta plain is characterised by channels flanked by well-defined levees and interchannel swamps with freshwater vegetation. The lower delta plain, within the intertidal zone, commonly shows a larger number of smaller channels affected by frequent overflow and shifting. Vegetation in this sector is tolerant to brackish and saline water. The delta front is the submerged sea-facing slope situated beyond the delta plain and the prodelta the platform across which the delta progrades. The morphology and dynamics of deltas largely depend on fluvial sediment input and the role played by waves and tides in redistributing and eroding the delta sediments. Fluvial-dominated deltas display elongate distributary branches projecting into the sea (e.g. Mississippi Delta) (Fig. 10.10A) or a lobate geometry (e.g. Nile Delta) (Fig. 10.10B). Deltas shaped by waves tend to have a cuspate geometry (e.g. Sao Francisco Delta, Brazil) (Fig. 10.10C). Those shaped by tides have an estuarine morphology with long inlets penetrating into the delta plain and elongated bars oriented perpendicularly to the shore direction (e.g. Ganges River). The majority of deltas in the world are experiencing severe degradation, largely due to reduced sediment input related to dam construction.

Fig. 10.3A A marine terrace consisting of raised beach deposits related to glacioisostatic rebound in Guinard Bay, northwest Scotland. In the background, roches moutonnées carved on Lewisian rocks (Published with permission of © Francisco Gutiérrez 2015. All Rights Reserved)

Fig. 10.3B A perched sloping shore platform cut across landward dipping limestone strata at Punta Nati, Menorca Island, Spain (Published with permission of © Francisco Gutiérrez 2015. All Rights Reserved)

10.11 Coral Reef

Coral reefs are the largest bioconstructional features on Earth and have a high preservation potential in the geomorphic and sedimentary record. They mostly develop in tropical areas with warm water. Other requirements for reef growth include good circulation, low nutrient content and clear water (low turbidity) that allows light penetration. Modern coral reefs have developed in the Holocene period, forming a relatively thin cover, typically less than 10–20 m thick, lying on a solu-

Fig. 10.3C Sloping marine strath (rasa) carved in Cretaceous limestone in Sonabia Bay, northern Spain. In the foreground, Sonabia beach at low tide and coastal dunes largely covered by vegetation (Published with permission of © Francisco Gutiérrez 2015. All Rights Reserved)

Fig. 10.4A Berria Beach, northern Spain, at low tide with a broad surf zone showing a bar and a trough in the foreground (Published with permission of © Francisco Gutiérrez 2015. All Rights Reserved)

tional–erosional unconformity carved during the lowstand of the last glaciation. Reefs in their juvenile stage grew vertically concomitantly with the sea-level rise. Once the reefs reach the modern sea level (mature stage), lateral accretion dominates, resulting in the development of reef flats and lagoons. In the senile stage, the lagoons are completely filled.

Reefs typically display a morphological and dynamic zonation parallel to the front. The reef crest is the highest fringe, exposed at low tide and with the strongest wave action. It is also called the algal ridge, since corals in this high-energy zone tend to be replaced by algae. The fore reef is the seaward slope, frequently bounded by a steep outer wall that

Fig. 10.4B Pocket sandy beach at Langre, northern Spain, in a bay surrounded by a cliff that culminates in a perched wave-cut platform cut across dipping strata (Published with permission of © Francisco Gutiérrez 2015. All Rights Reserved)

may reach more than 100 m in relief. The shallow and low-energy zone between the shore and the reef crest is the reef flat, which may include lagoons. Deposition in the lagoons is dominated by sand-sized particles largely derived from the erosion of the reef. The reefs attached to the shore of the mainland or islands are called fringing reefs. They are the most common reef type. Atolls are coral reefs found in open ocean comprising an annular rim that surrounds an inner lagoon (Fig. 10.11). Classical Darwinian atolls have grown upon volcanic edifices and may reach more than 1000 m in height, recording long-sustained subsidence of the oceanic crust, as it moves away from the spreading axis, and concomitant vertical reef accretion. There are also atolls in continental shelves mainly formed by coral debris and sand. A significant proportion of the coral reefs worldwide are undergoing human-induced degradation, mainly related to sedimentation, overexploitation and global warming. The rise in the water temperature has a detrimental effect on unicellular algae that play an instrumental role in corals' metabolism. Moreover, the increase in CO_2 concentration inhibits calcification by increasing the acidity of the water (Fig. 10.11).

Fig. 10.4C Gravel berm in the Cueva Beach, western Asturias, northern Spain, separating the beach face (left) and the landward sloping backshore. Published with permission of © Germán Flor 2015. All Rights Reserved

Fig. 10.4D Cusps in a gravel (shingle) beach at Cadavedo, western Asturias, northern Spain (Published with permission of © Germán Flor 2015. All Rights Reserved)

Fig. 10.5A Spit with a basalt stack partially closing a lagoon between Vik and Dyrholaey, southern Iceland. Published with permission of © Francisco Gutiérrez 2015. All Rights Reserved

Fig. 10.5B Spit with a sharp hook at its distal end in Provincetown, Barnstable County, Massachusetts. Image from © Google Earth

Fig. 10.5C Barrier evolved from a spit that has blocked a small drainage (Arroyo de la Laguna) creating a lagoon, Big Sur, California. Published with permission of © Francisco Gutiérrez 2015. All Rights Reserved

Fig. 10.5D Wash-over fan on Onslow beach barrier, North Carolina, after the 2012 Hurricane Sandy. Published with permission of © Beth van Dusen 2015. All Rights Reserved

Fig. 10.6A Lagoon at Sandwich Harbour south of Walvis Bay, Namibia, probably developed at the mouth of a paleodrainage defeated by dunes of the Namib Desert. Published with permission of © Francisco Gutiérrez 2015. All Rights Reserved

Fig. 10.6B Lesina Lagoon, eastern Italy, enclosed by a barrier related to the eastward growth of a spit from a headland situated on the western sector of the lagoon. Note the wash-over fans attached to the landward side of the barrier. An artificial channel was excavated on the western sector of the lagoon to improve the water exchange with the sea. Image from © Google Earth

Fig. 10.6C Sabkha around 20 km long at Al Lawi on the Red Sea coast, north of Jeddah, Saudi Arabia. Published with permission of © Francisco Gutiérrez 2015. All Rights Reserved

Fig. 10.7A A double tombolo in Orbetello, west coast of Italy, where the term tombolo comes from. Image from © Google Earth

Fig. 10.7B Tómbolo in Loch Eriboll, northern Scotland. Published with permission of © Javier Baztán 2015. All Rights Reserved

Fig. 10.8A Marshes and tidal flats in the Villaviciosa Estuary (Asturias, NW Spain). Published with permission of © Servicio de Cartografía del Principado de Asturias 2015. All Rights Reserved

Fig. 10.8B Tidal flat with an intricate system of channels in the estuary developed at the mouth of the Couesnon River, Normandy, France. The rock island on the left is Mont-Saint-Michel. Image from © Google Earth

Fig. 10.8C Marsh environments in the estuarine mouth of the Copper River, Cordova area, Alaska. Published with permission of © Francisco Gutiérrez 2015. All Rights Reserved

Fig. 10.8D Mudflats with tidal channels at the mouth of the Copper River, Cordova area, Alaska. Published with permission of © Francisco Gutiérrez 2015. All Rights Reserved

Fig. 10.9 Ria in the Adriatic coast of Montenegro, Skadarsko area. Published with permission of © Francisco Gutiérrez 2015. All Rights Reserved

Fig. 10.10A The fluvial-dominated Mississippi Delta with bird's-foot morphology. Image from © Bing Maps

Fig. 10.10B The Ebro Delta in eastern Spain, with well-developed spits on both flanks. The distal edge of the delta has retreated around 2 km since the 1960s, largely due to a major reduction in sediment flux induced by the construction of large dams. Image from © Bing Maps

Fig. 10.10C The São Francisco Delta in eastern Brazil is a typical wave-dominated cuspate delta. Image from © Google Earth

Fig. 10.11 Ebon Atoll around 13 km across developed upon a submerged volcanic seamount, Marshall Islands, Pacific Ocean. Image from © Google Earth

11 Glacial Landforms

Glaciers are ice masses that are thick enough to deform plastically and flow under their own weight (i.e. gravitational stress exceeds yield strength). They form mainly in high latitudes (polar regions) and highly elevated mountain environments, where snowfall is higher than snowmelt. As snow accumulates, it progressively increases in density transforming into firn (density 0.4–0.8 g/cm^3) and then into glacier ice (0.8–0.9 g/cm^3). This transformation includes a number of diagenetic processes including compaction, snow recrystallization and melting of snow followed by refreezing. The glacial budget is given by the balance between the overall rate of accumulation and the ablation rate of the ice. Glaciers with positive budgets increase their volume and advance, whereas those with negative budgets retreat. The temporal changes in ice mass extent are controlled by climatic factors, mainly temperature and precipitation. The equilibrium line defines the boundary between the upper accumulation and the lower ablation zones of glaciers. Glaciers, mainly as ice sheets, contain around 80 % of the world's freshwater and cover approximately 10 % of the emerged continental land surface. During the Last Glacial Maximum, at ca. 21,000 years ago, glaciers occupied around 30 % of the Earth's continental areas and the sea level was approximately 125 m lower than at the present time. The edge of the northern ice sheets reached latitudes as low as 40–50° (e.g. London is located at latitude 51°N). The largest was the Laurentide Ice Sheet, which created most of the landscape features of Canada and a significant portion of the northern USA. The weight of the ice sheets caused significant and extensive subsidence of the continental land masses by isostatic adjustment. Subsequently, the deglaciated areas have experienced progressive uplift (glacioisostatic rebound), as recorded by raised marine terraces and coastal cliffs.

The movement of glaciers is largely controlled by their subglacial thermal regime. Polar ice masses, in which the glacier bed is frozen to the ground (cold based), are dominated by plastic deformation. In contrast, basal sliding is the main displacement mechanism in temperate glaciers (warm based), where basal friction is reduced by liquid water, cavities and soft water-saturated sediments. These glaciers are more mobile and more effective in erosion. The flow in unconfined ice sheets is mainly determined by the slope of the ice mass, rather than the subglacial topography. The movement of confined valley glaciers is controlled by the topography, and they may overtop low interfluves through a process called diffluence.

The main processes involved in glacial erosion include the following: (1) abrasion, the wearing down of rock surfaces by debris-laden ice and meltwater; (2) plucking or quarrying, the detachment and removal of fractured rock fragments; (3) meltwater, which can reach extremely high velocities when it flows under hydrostatic pressure; and (4) freeze–thaw, mainly in temperate glacial environments. Debris may be transported by ice or meltwater in supraglacial, englacial, subglacial and proglacial environments. Sediments deposited by the ice mass, commonly referred to as till or glacial diamicton, are typically massive and unsorted, whereas sediments accumulated by flowing water (fluvioglacial) or related to proglacial water bodies (glaciolacustrine, glaciomarine) are generally sorted, finer grained and stratified.

In the current global warming situation, glacial dynamics has relevant practical implications. The progressive melting of the ice masses is causing a progressive sea-level rise with potentially alarming flooding consequences for low coastal areas (e.g. Bangladesh). Icebergs largely derived from unstable ice-sheet margins have detrimental effects on navigation and oil rigs. Conversely, glacial retreat may offer development opportunities in areas formerly covered by ice, and the decrease in the Arctic sea ice will allow new shipping routes. Ice avalanches constitute a significant hazard in some mountain regions like the Himalayas. The largest-ever recorded floods on Earth are related to the outburst of ice-dammed lakes, which may form either by blocked drainage or by subglacial volcanic eruptions resulting in glacial meltwater outbursts (jökulhlaups).

F. Gutiérrez, M. Gutiérrez, *Landforms of the Earth*, DOI 10.1007/978-3-319-26947-4_11

11.1 Glacier

The largest and thicker glaciers are the ice sheets. The ice sheets of Antarctica and Greenland represent more than 95 % of the world's glacierized surface. Ice sheets are characterised by a dome-like topography and a general outward flow, barely controlled by the buried subglacial landscape due to the high thickness of the ice. The maximum thickness of ice in Antarctica and Greenland is around 4.7 km and 3 km, respectively. The term ice cap is used for thick, dome-shaped glaciers less than 50,000 km^2 mostly developed in elevated areas (Fig. 11.1A). Ice fields are ice masses that commonly form on elevated plateaus, have lower thickness and extent than ice caps and may display an undulating topography that reflects the subglacial bed. Ice sheets, ice caps and ice fields may be locally interrupted by protruding mountains surrounded by ice that escape glacial erosion called nunataks (Fig. 11.1B). At the margins of these glaciers, the ice may radiate along erosional gaps forming tongue-shaped outlet glaciers (Fig. 11.1C). The marginal sectors of ice sheets and ice caps may float in the ocean forming extensive slablike ice shelves. In alpine or mountain glaciers, the morphology of the ice mass and its flow is governed by the topography. Valley glaciers are elongated ice streams confined by the sides of a valley (Fig. 11.1D). When the lower part of a valley glacier leaves the mountain terrain and reaches a plain, it expands laterally to form lobe-shaped piedmont glaciers (Figs. 11.1G and 11.1K). Cirque glaciers are small glaciers restricted to the commonly steep-sided depressions carved in valley heads and mountain sides. These ice masses typically have a low length/width ratio (Figs. 11.1E and 11.1F).

The movement of glaciers, strongly controlled by the temperature of basal ice, is mainly related to plastic deformation and basal sliding. Cold-based glaciers, with basal ice frozen to the ground, are dominated by plastic deformation. In contrast, movement in temperate glaciers with lubricating liquid water at their base is mainly achieved by basal sliding. Differential displacement within the ice mass produces a wide variety of deformation structures like crevasses (Figs. 11.1D and 11.1H). These are extensional fissures that commonly occur as systems with a consistent orientation. Transverse crevasses are generally associated with extending flow (downslope acceleration of the flow) and splaying crevasses with compressional zones (ice flow decelerates downslope). Oblique crevasses are common along the lateral margins affected by shearing between the ice and the valley walls. The frequently large fissure that forms in the glacier head, between the cirque wall and the ice moving downslope, is called bergschrund (Fig. 11.1I). Large ice blocks may detach or fall from the edge of glaciers associated with the sea, lakes or rivers, producing icebergs by a process called calving (Fig. 11.1J).

11.2 Cirque

Cirque is a French word widely used to designate rock basins carved by glacial erosion on mountain sides. These hollows, typically hundreds of metres deep and wide, are partially enclosed by an arcuate rock cliff (headwall) (Fig. 11.2A). The floor frequently shows one or several closed depressions and the outer edge may correspond to a bedrock threshold. The overdeepened basins are frequently occupied by lakes (Fig. 11.2B). These hollows commonly hang above glacial troughs and are best developed on slopes with favourable aspect for the accumulation of snow, which may depend on factors like the prevailing wind direction or solar insolation (e.g. lower in N–NE-facing slopes in the northern hemisphere). The spatial distribution of cirques seems to be related to that of pre-existing concavities adequate for the development of nivation hollows that evolve into cirques. The formation and enlargement of cirques is related to the

Fig. 11.1A The dome-shaped Myrdalsjökull ice cap in southern Iceland covers approximately 600 km^2 and is underlain by the Katla volcano, whose last eruption occurred in 1918 producing a major jökulhlaup (Published with permission of © Francisco Gutiérrez 2015. All Rights Reserved)

Fig. 11.1B Air view of nunataks in southeastern Greenland (Published with permission of © Nicolas Benoist 2015. All Rights Reserved)

Fig. 11.1C Fjallsjökull is an outlet glacier of the Vatnajökull ice cap in southern Iceland. The glacier terminates at a proglacial lake with icebergs, located behind a frontal moraine that records the rapid retreat of the glacier in historical times (Published with permission of © Francisco Gutiérrez 2015. All Rights Reserved)

progressive deepening of the floor and the retreat of the headwalls. Glacial deepening is mainly accomplished by plucking and abrasion intensified by the displacement of the ice following a rotation trajectory (rotational basal sliding). Headwall retreat is primarily related to rock fall and rock avalanche processes from the oversteepened cliffs. Headwall retreat in adjacent cirques may result in the development of sharp ridges called arêtes. Arêtes related to several cirques may converge in pyramidal peaks, also called horns (Figs. 11.2C and 11.2D).

11.3 Glacial Valley and Fjord

Most glacial valleys are generated by ice streams channelled along pre-existing fluvial valleys. Glacial erosion progressively leads to the overdeepening and widening of the valley. Glacial valleys frequently show U-shaped transverse profile with steep slopes, especially where carved in resistant bedrock (Fig. 11.3A). In contrast to fluvial valleys, the longitudinal profile of glacial valleys typically shows an undulating geometry, including closed basins, rock steps and reaches with reverse gradient. The location of overdeepened basins may be related to several factors like bedrock resistance, the incorporation of tributary ice streams or valley constrictions. These enclosed basins in the valley floor are initially occupied by lakes that fill progressively by lacustrine and fluvial deposits until they become flat swampy peat bogs. Hanging valleys are tributary glacial troughs that end abruptly at the steep side of a glacial valley with higher entity, where glacial overdeepening has been much more intense. These sharp knickpoints are commonly the locus of spectacular waterfalls (Fig.11.3B). The sides of valley glaciers, especially those that have been recently deglaciated, may show a conspicuous line that indicates the upper edge of the glacially eroded surface called trimline (Fig. 11.3C). Fjord is a Norwegian word used to describe glacial troughs submerged by the sea. This situation is largely related to the sea-level rise resulting from the melting of the Pleistocene ice sheets during the last interglacial (Fig. 11.3D).

11.4 Roche Moutonnée

Roche moutonée is a French term that designates asymmetric rock hills moulded by glacial erosion. They may form extensive fields and display a wide range of dimensions, but typically reach tens of metres high and hundreds of metres long. These bumps are characterised by smooth upstream sides and steeper and frequently irregular downstream faces. This asymmetry seems to reflect the relative importance of two different erosion processes: abrasion on sides facing upflow and plucking on the opposite sides. Where the ice flows against an obstacle, it tends to smooth its upstream side by abrasion reducing the frictional resistance. This process is caused by debris embedded in and dragged by the ice under pressure. Abrasion produces a number of small erosional features including polished surfaces, stream-lined striations and grooves, as well as transverse crescentic fractures. On the downstream side, where the pressure is lower, the ice tends to detach and entrain blocks from the jointed bedrock by plucking. Moreover, higher pressure in the upstream side may induce ice melting favouring basal sliding and abrasion; an increase in pressure is accompanied by a decrease in the melting temperature (pressure melting point). Conversely, on the downstream side of the obstacle, where the pressure is lower, the meltwater may refreeze and weaken the rock mass by frost shattering promoting plucking. These glacial erosion processes are more intense in temperate glaciers with significant basal sliding favoured by the presence of liquid water at the base (Figs. 11.4A and 11.4B).

Fig. 11.1D Skaftafellsjökull is a valley and outlet glacier sourced in the Vatnajökull ice cap. Note the medial moraine and the crevasses with a dominant oblique orientation (Published with permission of © Francisco Gutiérrez 2015. All Rights Reserved)

Fig. 11.1E The Aneto (left) and Maladeta (right) cirque glaciers are restricted to cirques carved in the Maladeta granitic massif, central Spanish Pyrenees (Published with permission of © José María García-Ruiz 2015. All Rights Reserved)

Fig. 11.1F The Monte Perdido cirque glacier comprises two ice bodies lying on structurally controlled benches, Ordesa and Monte Perdido National Park, Spanish Pyrenees (Image taken in July 1995. Published with permission of © Francisco Gutiérrez 2015. All Rights Reserved)

11.5 Moraine

Glaciers transport debris along its base (subglacial), within the ice (englacial), on the surface (supraglacial) and at their edges (ice marginal). Moraine refers to the landforms created by the deposition or deformation of sediments by glaciers. Till is a broad term used to designate morainic deposits typically consisting of unstratified and poorly sorted sediments. Moraines can be classified according to their position and formation mechanism. The generation of moraines can be related to the release of debris by ablation, deformation of subglacial or proglacial sediments by the ice move-

Fig. 11.1G Outlet glaciers from ice caps that expand in the lower unconfined part to form piedmont glaciers, Axel Heiberg Island, Canadian Arctic (Image from © Bing Maps 2015)

ment and the accumulation of deposits along the edge of the glaciers. Lateral moraines form along the flanks of valley glaciers and consist of debris derived from the valley walls, both above and beneath the ice. They may show multiple nested ridges related to different recession stages and may dam tributary valleys forming moraine-dammed lakes (Fig. 11.5A). Medial moraines are longitudinal accumulations that form along the edge of two confluent ice streams downstream of their junction (Fig. 11.5B). Frontal moraines develop along the downstream edge of glaciers. These are ridges that commonly display arcuate geometry in plan view and may connect with lateral moraines (Fig. 11.5C). Moraine-dammed lakes are also common upstream of frontal moraines (Fig. 11.5C). The formation of frontal moraines may be related to the bulldozing of sediment at the front of an advancing glacier (push moraines) or the release of supraglacial and englacial debris by ablation in a glacier with a stable (dump moraine) or retreating edge (ablation moraines). This latter process may produce relatively thick accumulations with an irregular topography called hummocky moraines (Fig. 11.5C). Push moraines have an internal structure with downstream-verging thrusts and folds and may show multiple ridges related to seasonal bulldozing of proglacial deposits (Fig. 11.5D). The term terminal moraine is used to indicate the morainic ridges that record the maximum extent of the glaciers. Subglacial moraines may show irregular topography without any preferred trend or linear features with a consistent orientation. The accumulations of subglacial till without any distinctive geomorphic pattern are commonly described as ground moraine. Rogen or ribbed moraines are long ridges transverse to ice flow tens of metres high that commonly display a slight concavity facing upstream. These subglacial moraines may cover extensive areas and seem to be related to deformation of subglacial sediments or debris-rich basal ice. Ablation moraines, especially those related to ablation, include isolated portions of buried ice called dead ice (Fig. 11.5E). The melting of these ice bodies leads to the subsidence of the overlying deposits (thermokarst) and the formation of enclosed depressions known as kettle holes. Erratics are isolated blocks left on the ground by retreating glaciers (Fig. 11.5F). The lithology of these allocthonous blocks, frequently different to that of the underlying rocks, may be used to infer the ice flow patterns. Similarly, moraines are useful geomorphic markers for reconstructing the extent and paleogeographical evolution of glaciers in the past.

Fig. 11.1H Transverse crevasses in the upper part of the Aneto Glacier, Spanish Pyrenees (Published with permission of © Francisco Gutiérrez 2015. All Rights Reserved)

11.6 Drumlin

The word drumlin derives from a Gaelic term (*droimnín*) that refers to mounds. Drumlins are elongated rounded hills mainly composed of glacial deposits formed in subglacial environments and with the major axis oriented in the direction of the ice flow. Although they may display a wide variety of geometries, the archetypal drumlin has an inverted spoon-like shape with steeper upstream and tapering downstream slopes. They may reach 200 m in height and several kilometres in length, and typically occur as extensive fields, which may have the appearance of a "basket of eggs". They may have a very complex internal composition, including glacial till, stratified and well-sorted glaciofluvial facies and in some cases a bedrock core. Downflow-verging glaciotectonic structures are common. Drumlins mostly occur in areas formerly covered by ice sheets and broad valley glaciers. The origin of these enigmatic landforms has received two main interpretations. One of them is the remoulding of subglacial deposits by the moving ice. According to the other interpretation, drumlins are related to extreme subglacial meltwater flows (mega-floods). These rapid flows may either carve cavities that are subsequently filled by deposits or may sculpt drumlins by eroding pre-existing deposits. The distribution of these two processes may be favoured by the presence of resistant bedrock bumps. Most current studies into drumlin genesis are employing the concept that anisotropies in basal ice and/or substrate lead to drumlinization of the glacier bed (Figs. 11.6A, 11.6B, and 11.6C).

Fig. 11.1l Bergschrung developed in the upper edge of the Aneto Glacier and at the foot of the cirque headwall (Published with permission of © Francisco Gutiérrez 2015. All Rights Reserved)

11.7 Esker

The name Esker is derived from the Irish word "eiscir" meaning ridge or elevation separating two plains or depressed surfaces. Eskers are long, narrow and sinuous ridges composed of glaciofluvial sediments deposited in subglacial, englacial and supraglacial channels. The subglacial and englacial channels typically correspond to tunnels that may form a well-integrated system of conduits like those found in karst aquifers. Supraglacial channels may connect with subsurface channels through shafts. When the outflow capacity of the "plumbing system" is overwhelmed by water inflow, water under high pressure in the subsurface channels may flow at extremely high velocities, move upwards over obstacles and transport large boulders common is eskers. These situations may be related to substantial incorporation of meltwater, as may occur at the end of hot days in summer, or in relation to severe rainfall events. Eskers may occur as single ridges or as braided or distributary complexes. They may be continuous and segmented or consist of a closely spaced succession of mounds (beaded). They are typically oriented in the direction of the ice flow and may reach hundreds of kilometres in length and tens of metres in height. Eskers formed in subglacial meltwater channels are found where they were initially constructed, often splaying out into glaciofluvial or glacio-

Fig. 11.1J Icebergs in Jökullsarlon Lake in southern Iceland. This lagoon has formed behind a frontal moraine of the Breidamerkurjökull outlet glacier (background) of the Vatnajökull ice cap, Iceland. Note the perched notches in the icebergs and the debris-rich beds mainly related to pyroclastic deposits. Medial moraines and crevasses in the outlet glacier that produces the icebergs (Published with permission of © Francisco Gutiérrez 2015. All Rights Reserved)

lacustrine fans at their downstream end, whereas the deposits of eskers related to supraglacial or englacial channels have settled during glacier melting to drape the ground surface. Esker sediments commonly show sorted, rounded and stratified sand and gravel facies related to multiple water flow events. Boulder and cobble gravel beds indicative of a high-energy environment are common. An internal arched structure with beds dipping away from the axis is often related to collapse of the confining ice walls during the ice mass decay (Figs. 11.7A and 11.7B).

11.8 Outwash Plain

Outwash plains, also designated as sandur, are glaciofluvial environments developed downstream of glaciers and mainly fed by meltwater. These are typically gravelly braided systems that may be confined to the bottom of a valley or occur on single or coalescing fans with distributary channels (Fig. 11.8A). The water flow in outwash plains has a strong seasonal and diurnal pattern and a marked thermal control, since most of the meltwater is discharged during a short summer period, chiefly during daytime. Outwash plains may be affected by catastrophic floods called jökulhlaups, which may be related to the sudden drainage of ice-dammed lakes or subglacial volcanic activity. In both types of floods, the water pressure may be high enough to force its way under the ice. These floods may deposit large amounts of debris in the outwash plains, including ice blocks easily transported as floating particles. The progressive melting of the buried ice leads to the development of subsidence depressions known as kettle holes. Kettle holes may also develop in the proximal sector of outwash plain, next to the ice front, where glaciofluvial sediments may develop over stagnant ice (Fig. 11.8B).

Fig. 11.1K Bering piedmont glacier and Vitus proglacial lake. Note the folding structure developed in the frontal compressional zone of the glacier. The glacier terminus has retreated more than 12 km since 1900 despite the frequent surging events. Lateral moraine complex with kettle holes on the *left* (Image from © Bing Maps 2015)

Fig. 11.2A Salenques cirque with well-defined headwall carved in granitoids of the Maladeta Massif, central Spanish Pyrenees (Published with permission of © Francisco Gutiérrez 2015. All Rights Reserved)

Fig. 11.2B Complex cirque with multiple overdeepened basins excavated in a granitic batholith. Note strongly deformed sedimentary rocks affected by contact metamorphism, including white marble. Infierno, Arnales and Bachimaña peaks and Azules glacial lakes, Panticosa, Spanish Pyrenees (Published with permission of © Francisco Gutiérrez 2015. All Rights Reserved)

Fig. 11.2C Arêtes developed along the jagged ridges that separate adjacent cirques in the granitic Maladeta Massif, Spanish Pyrenees (Published with permission of © Francisco Gutiérrez 2015. All Rights Reserved)

Fig. 11.2D The Matterhorn (Monte Cervino) is the classical horn, located at the border between Switzerland and Italy in the Alps (Published with permission of © Márcio Décio 2015. All Rights Reserved)

Fig. 11.3A The steep-sided Ordesa glacial valley, Spanish Pyrenees, has been carved in a thick limestone and sandstone succession related to the stacking of thrust sheets (Published with permission of © Francisco Gutiérrez 2015. All Rights Reserved)

Fig. 11.3B Hanging glacial valley and the ca. 190 m high Bridalveil Waterfall in Yosemite National Park, California (Published with permission of © Francisco Gutiérrez 2015. All Rights Reserved)

Fig. 11.3C Trimline related to the rapid retreat of the Gulkana Glacier that emanates from an ice field in the eastern Alaska Range (Published with permission of © Francisco Gutiérrez 2015. All Rights Reserved)

Fig. 11.3D Seydisfjördur fjord in NE Iceland is a glacial valley partially submerged by the sea (Published with permission of © Francisco Gutiérrez 2015. All Rights Reserved)

Fig. 11.4A A roche moutonnée carved by a Pleistocene glacier in granite with a steep downflow side and a gentler upflow side, Tuolumne Meadows, Yosemite National Park, California (Published with permission of © Francisco Gutiérrez 2015. All Rights Reserved)

Fig. 11.4B Small stream-lined roche moutonnéess carved in a bedrock threshold by the Flaajökull glacier in southern Iceland. Note the smooth abraded upflow side with striations and the steep lee side affected by joint controlled plucking (Published with permission of © Francisco Gutiérrez 2015. All Rights Reserved)

Fig. 11.5A Pleistocene lateral moraine in the Miera Valley, Cantabria, northern Spain. The adjacent sediment-filled glaciolacustrine basins are related to the damming of tributary drainages by the moraine (Published with permission of © Francisco Gutiérrez 2015. All Rights Reserved)

Fig. 11.5B Medial moraines in Gulkana Glacier, eastern Alaska Range (Published with permission of © Francisco Gutiérrez 2015. All Rights Reserved)

Fig. 11.5C Prominent ridge corresponding to a lateral–frontal moraine of the Kviarsjökull glacier in southern Iceland. The undulating ground with kettle holes to the right of the proglacial lake corresponds to a meltout hummocky moraine related to the rapid historical recession of the glacier (Published with permission of © Francisco Gutiérrez 2015. All Rights Reserved)

Fig. 11.5D Multiple seasonal push frontal moraines of the Skalafellsjökull glacier, Iceland (Published with permission of © Francisco Gutiérrez 2015. All Rights Reserved)

Fig. 11.5E Debris-rich dead ice at the front of the Solheimajokull glacier, Iceland. The conical mounds in the glacier snout are related to locations of decelerated ablation caused by the insulating effect of a debris cover (Published with permission of © Francisco Gutiérrez 2015. All Rights Reserved)

Fig. 11.5F Erratic block in the Marboré cirque, central Spanish Pyrenees. Monte Perdido glacier in the background (Published with permission of © Francisco Gutiérrez 2015. All Rights Reserved)

Fig. 11.6A Tree-covered drumlin west of Geneva, Ontario County, New York. The pointed edge of the drumlin indicates the direction of the formative ice flow (Image from © Bing Maps 2015)

Fig. 11.6B Drumlin field in Sligo area, Ireland. The ice flow is to the left (Image from © Bing Maps 2015)

Fig. 11.6C Ground view of drumlin east of Sligo Town, northwest Ireland. Ice flow towards the *left* (westwards) (Published with permission of © Xavi Mir 2015. All Rights Reserved)

Fig. 11.7A Esker (linear island) in Rorstromssjon, northwest Angermanland, Sweden (Image from © Bing Maps 2015)

Fig. 11.7B Split Hill Esker located north of Tullamore Town, Irish Midlands (Photograph pointing downstream. Copyright Xavi Mir)

Fig. 11.8A Outwash plain (sandur) with an unconfined gravelly braided system in front of Skaftafellsjökull outlet glacier, southern Iceland. The ponded closed depressions on the *left* correspond to kettle holes (Published with permission of © Francisco Gutiérrez 2015. All Rights Reserved)

Fig. 11.8B Fresh kettle holes related to thawing of buried ice blocks deposited by the 1996 jökulhlaup developed in Skeiðarársandur, Iceland (Published with permission of © Zbigniew Zwolinsli 2015. All Rights Reserved)

Periglacial Landforms

12

Periglacial environments occur in cold non-glacial regions characterised by intense freezing and thawing activity and perennially or seasonally frozen ground. They occur in zones where the mean annual air temperature is lower than 3 °C and are not restricted to ice-marginal regions, as their name suggests. Periglacial environments occupy around 20 % of the world's continental area and have a broad latitudinal distribution, comprising extensive high-latitude tundra and taiga regions (e.g. Siberia, Canada, Alaska) and numerous high-elevation mountain and plateau regions (e.g. Tibet) with a broad latitudinal distribution. Permafrost refers to the ground (soil or rock) that remains frozen for more than 2 years. The near-surface zone that thaws in summer is called the active layer. The underlying permafrost may be either continuous or discontinuous where the frozen ground is locally interrupted by patches of unfrozen ground called talik.

Periglacial morphoclimatic regions show distinctive landforms and deposits related to intense frost action, ice thawing and the presence of frozen ground. The aggradation of the permafrost causes progressive uplift of the ground surface. The growth of massive or lenticular ice bodies (ice segregation) may produce mounds such as pingos or palsas. In contrast, the thawing of ground ice leads to ground settlement and the development of thermokarst subsidence depressions. Ice-rich detrital deposits on hillsides may flow downslope by plastic deformation resulting in the development of rock glaciers. In relatively flat surfaces underlain by unconsolidated deposits, thermal contraction induced by temperature lowering can generate polygonal cracks. The opening of these cracks may be enhanced by desiccation of unfrozen water and the fissures may be filled by ice wedges. Aggradation of ice in the wedges results in lateral pressures and the bulging up of the adjacent ground.

Frost shattering or congelifraction of rocks at or near the surface typically disintegrates them into angular clasts. This is typically more intense in regions with numerous freeze–thaw cycles. The fragmentation of the rock mass may be caused by the combined effect of several processes, including the volumetric expansion of the water during freezing, hydrofracturing caused by local high pore-fluid pressures induced by rapid propagation of freezing fronts in undrained conditions, growth of ice lenses as water migrates towards a freezing front and differential volume changes in the rock minerals related to temperature variations. Persistent snow patches are common in mountain periglacial regions. In those settings a number of processes are intensified by snowmelt water, favouring differential erosion (nivation).

Periglacial conditions have important environmental and engineering implications, the future impacts of which are expected to increase driven by global warming. The permafrost poses difficult ground conditions for construction, especially when the thermal equilibrium is disturbed by human activity (e.g. vegetation clearance, engineering structures, impoundment of reservoirs). The thawing of the permafrost may cause significant ground subsidence resulting in severe damage to human structures. Frozen ground also complicates the exploitation and transportation of mineral resources, including hydrocarbons. Pipelines used for the conveyance of oil and gas require special costly designs to insulate them from the ground and accommodate potential deformation related to freezing and thawing processes. Seasonal thawing commonly involves a substantial reduction in the mechanical strength of the active layer, favouring the development of slope movements. The hydrology of a significant proportion of periglacial areas is largely controlled by the annual thermal regime. During the long winters, rivers and lakes may be covered by ice. In the summer, thick snow packs may rapidly melt producing severe floods.

12.1 Talus Slope and Talus Cone

Talus and its synonymous scree are used to designate deposits of loose, coarse and generally angular clasts accumulated at the foot of a rock cliff. They typically form thick wedge-shaped accumulations with a straight longitudinal profile and a basal concavity that grow by parallel accretion. The slope

F. Gutiérrez, M. Gutiérrez, *Landforms of the Earth*, DOI 10.1007/978-3-319-26947-4_12

angle, determined by the friction angle of the cohesionless debris, is generally between 30° and 40°. This type of depositional landform occurs in a wide range of environmental conditions, but is particularly common in mountainous periglacial environments. Frost shattering (congelifraction) plays a major role in the production of debris from the rock cliff mainly through rock falls. The fallen blocks move downslope bouncing, rolling and sliding until they halt. Debris flows, sediment-laden water flows and debris-bearing snow avalanches may also have a significant contribution. Talus typically shows openwork texture, with empty interclast spaces, although the amount of matrix may increase with depth due to downwashing of fines by percolating water. The high permeability of these deposits hinders surface run-off and consequently reduces their vulnerability to water erosion. Elongated and planar clasts tend to display prevalent fabrics concordant with the profile of the slope. Talus also shows longitudinal sorting, with the largest blocks in the distal sectors. Large blocks have higher momentum and are affected by proportionally lower friction due to their low specific surface. Talus slopes have a planar geometry and form at the foot of non-dissected cliffs (Fig. 12.1A). Talus cones have convex cross profile and a well-defined apex connected to an incision in the cliff (couloir). Those channels tend to funnel and confine rock falls, surface run-off, debris flows and snow avalanches (Fig. 12.1B).

12.2 Block Field, Block Slope and Block Stream

Congelifraction or frost shattering in bedrock involves the fragmentation of the rock by the freezing of infiltrated water and segregated ice in joints and interstitial pores. When this process operates upon hard rocks in which the discontinuity planes have a relatively wide spacing, it produces boulder-sized angular blocks. Block fields or felsenmeer are widespread mantles of angular blocks with openwork texture and no visible fine material that occur on nearly flat surfaces (Fig. 12.2A). They frequently develop on hard rocks with siliceous composition that produce limited amount of fines upon weathering (e.g. granitoids, gneiss, quartzite). Block fields are particularly common in planation summits of mid-latitude mountain ranges. They are essentially an autochthonous breccia with a transitional contact with the undisturbed bedrock. The rubble commonly shows a progressively higher

Fig. 12.1A Talus slope developed below a rock cliff underlain by sandstones, conglomerates and andesites. The cliff supplies blocks with a wide range of sizes and larger blocks tend to accumulate in the distal sectors (Barranco de Cubillas, Tena Valley, Spanish Pyrenees) (Published with permission of © Francisco Gutiérrez 2015. All Rights Reserved)

Fig. 12.1B Debris cones nested behind moraines and fed by steep troughs carved into a north-facing limestone escarpment. Debris flows play a major role in the building up of the cones, as reveal the levees and lobate accumulations on their surface (Partacua Range, Spanish Pyrenees) (Published with permission of © Francisco Gutiérrez 2015. All Rights Reserved)

amount of relative displacement towards the surface, showing a gradation from crackle to mosaic and chaotic breccia. In areas where the slope angle is high enough, the blocks may undergo some transport mainly by mass movement processes to form block slopes (Fig. 12.2B). The clasts in the slopes commonly show vertical sorting with larger particles at the top due to frost heaving and frost sorting processes. They also display higher proportion of fines downwashed by percolating waters and throughflow towards the base of the deposit and the lower sector of the slope. Block slopes from hillsides may merge in gullies and valley floors to form block streams (Fig. 12.2C). These deposits that may reach some kilometres in length and more than 10 m in thickness tend to fill a pre-existing dissected topography. Block streams also show openwork texture at the surface, reverse grading (downward fining) and some fine-grained matrix in the basal part. Block slopes and block streams may show lobes and benches related to solifluction. This transport mechanism may play a significant role on the displacement of the blocks when they accumulate on argillaceous bedrock, behaving as a passive raft transported over an underlying slow flow-type slope movement. The boulders of inactive block streams typically show rounded edges and a cover of lichens and mosses. These relict landforms, although difficult to date, may constitute valuable paleoenvironmental indicators.

12.3 Grèzes Litée

Grèzes litées is the French term used to describe rhythmically stratified slope deposits consisting of alternating beds of angular gravel, mostly pebble-sized and fine-grained deposits. They display parallel bedding concordant with the slope, centimetre- to decimetre-scale layer thickness, and the gravelly beds are commonly moderately sorted and show a matrix-free openwork texture. Grèzes litées are mainly found below rock cliffs in midlatitude periglacial environments with a high number of freeze–thaw cycles and are apparently absent in high latitudes. They are particularly well developed in slopes underlain by well-bedded limestone. The resulting calcareous deposits are frequently cemented by dissolution and precipitation processes related to percolating water. There is wide consensus about the decisive role played by frost shattering caused by a high number of freeze–thaw cycles in the formation of these deposits. However, the origin of the stratification, which indicates alternating conditions,

Fig. 12.2A Block field produced by frost shattering of Proterozoic granite in Trail Ridge, Rocky Mountain National Park, Colorado (Published with permission of © Francisco Gutiérrez 2015. All Rights Reserved)

and the rhythmic alternation of coarse- and fine-grained beds, is still an unresolved debate. Different processes including snow melting, slope wash, solifluction and frost heave, among others, have been claimed to explain these features (Fig. 12.3).

12.4 Nivation Hollow and Protalus Rampart

Nivation hollows are erosional depressions generated in slopes by intensified weathering and erosion associated with lingering snow patches. These snowdrifts typically occur in locations with adequate topography and favourable aspect with respect to solar radiation and wind, which redistributes and preferentially accumulates the snow in topographic traps. They are characterised by a steep backslope and a low relief and gently sloping floor. Nivation hollows are considered by some authors as the embryonic stage of glacial cirques. Nivation collectively refers to a group of processes associated with, or intensified by the presence of persistent snowdrifts, which provides snowmelt water as the main weathering and erosion agent. Snow patches also hinder the development of a protective vegetation cover. The formative processes include enhanced frost shattering along the margins and beneath the snow cover, slope wash, mass wasting such as solifluction and chemical weathering. Protalus ramparts are ridges of debris deposited along the downslope margin of snow patches, commonly situated below a steep bedrock slope. Rockfall debris from the cliff, largely related to frost shattering, moves downslope over the snow rolling, bouncing and sliding to preferentially accumulate at the foot of the snow-covered slope. Other processes such as debris-bearing snow avalanches and debris flows may also contribute to the construction of the ridge. The ridges may be linear or sinuous and frequently show asymmetric profiles, with a steeper and shorter proximal slope. The grain size of the deposit largely depends on the characteristics of the bedrock in the backwall. They typically consist of unsorted gravels and boulders with chaotic fabrics and may include fines related to in situ weathering and rock disintegration by impacting clasts. Protalus ramparts may occur as a series of parallel ridges attributable to both snow pack and backwall recession (Figs. 12.4A, 12.4B, and 12.4C).

Fig. 12.2B Block slope develop on rhyolite bedrock, Cerler Peak, Benasque Valley, Spanish Pyrenees (Published with permission of © Francisco Gutiérrez 2015. All Rights Reserved)

12.5 Rock Glacier

Rock glaciers are one of the most remarkable periglacial features of mountain areas with current or past permafrost. Active rock glaciers are defined as lobate or tongue-shaped bodies of detrital material with internal ice that flow downslope due to the gravity-driven plastic deformation of the ice contained in them. These deposits are generally found at the foot of steep talus slopes in glacial cirques and valleys. They typically consist of an upper blocky layer underlain by fine-grained material supersaturated with ice. Rock glaciers derived from the decay of debris-covered glaciers may have a core of glacial ice. These flow-like slope movements are characterised by steep frontal and lateral slopes and a corrugated surface of arcuate ridges and furrows with a general downslope convexity. They can be classified as talus rock glaciers or debris rock glaciers depending on whether the deposits are derived from talus slopes or moraines, respectively. Lobate and tongue-shaped rock glaciers have length to width ratios lower and higher than one, respectively. Spatulate rock glaciers have a widened lower part. As the internal ice of rock glaciers melts, they stop moving downslope and acquire a subdued morphology with thermokarst subsidence features. These relict rock glaciers may provide valuable information on past permafrost and paleoclimatic conditions (Figs. 12.5A, 12.5B, and 12.5C).

Fig. 12.2C Relict block stream consisting of quartzite blocks with rounded edges and a weathering rind that supports a lichen cover, Orihuela del Tremedal Massif, Iberian Chain, Spain (Published with permission of © Francisco Gutiérrez 2015. All Rights Reserved)

Fig. 12.3 Relict grèzes litées exposed in an aggregate quarry at the foot of a limestone cliff in Las Parras River valley, Iberian Chain, NE Spain (Published with permission of © Francisco Gutiérrez 2015. All Rights Reserved)

Fig. 12.4A Nivation hollow with protalus rampart, Bejar Range, Central System, Spain (Published with permission of © Rosa Carrasco 2015. All Rights Reserved)

Fig. 12.4B Arcuated protalus rampart at the foot of a rock cliff on slates with interbedded quartzites. The ridge has a curved geometry and an asymmetric cross profile. The high content of fines is mainly related to rapid in situ disintegration of slate clasts (Remáscaro catchment, Benasque Valley, Spanish Pyrenees) (Published with permission of © Francisco Gutiérrez 2015. All Rights Reserved)

Fig. 12.4C Protalus rampart consisting of openwork angular boulders of metamorphic rocks at the downslope edge of a late-lying snow pack (southern flank of the Garmo Negro Peak, Panticosa, Spanish Pyrenees) (Published with permission of © Francisco Gutiérrez 2015. All Rights Reserved)

Fig. 12.5A Tongue-shaped rock glacier nested behind a turf-covered neoglacial moraine and consisting of gneiss blocks, Pacific Peak, Breckenridge, Southern Rocky Mountains, Colorado (Published with permission of © Francisco Gutiérrez 2015. All Rights Reserved)

Fig. 12.5B Lobate rock glaciers generated from talus slopes consisting of gneiss blocks, Mohawk Cirque, Breckenridge, Southern Rocky Mountains, Colorado (Published with permission of © Francisco Gutiérrez 2015. All Rights Reserved)

Fig. 12.5C Tongue-shaped rock glacier around 2 km long in Mount Sopris, Southern Rocky Mountains of Colorado (Image from ©Bing Maps)

12.6 Patterned Ground

Patterned grounds are geometrical features commonly arranged in a regular pattern that mainly form in unconsolidated sediments within the active layer of permafrost areas or in seasonal frozen ground. Similar patterned ground develops in other climatic environments as a result of a wide range of processes, like gilgai related to swelling and shrinkage of clays in dry areas (morphologic convergence or equifinality). Patterned ground in periglacial areas can be described according to their geometry (circles, polygons, nets, stripes and steps) and depending on whether it is sorted or non-sorted. The term net is used where circles or polygons form a mesh of interacting cells. Circles, polygons and nets occur on subhorizontal surfaces (Figs. 12.6A and 12.6B). Stripes and steps develop on sloping surfaces and may show a gradation to polygons and circles with decreasing gradient (Figs. 12.6C, 12.6D, and 12.6E). The origin of patterned ground is attributed to a wide range of processes (polygenetic landform): Thermal contraction cracking: lowering of the temperature causes contraction in ice-rich frozen soil and the formation of cracks. Desiccation cracking: fissures may develop by water evaporation in unfrozen ground. Ice wedging: ice wedges may form and aggrade in the cracks pushing laterally the adjacent sediment. Frost heaving: freezing of moisture in the active layer causes the upward expansion of the soils. Frost sorting: particles in heterogeneous deposits tend to be segregated by differential frost heave and thaw settlement. For instance, larger particles accumulate at the edges of bulging polygons or circles. Thaw settlement: thawing of ice may lead to contraction and subtle ground settlement. Patterned ground commonly has a decimetre or metre scale. However, polygonal ice wedge nets with raised rims and diameters of the order of tens of metres cover extensive tracts in Arctic areas. The most common pattern is the non-sorted circle, termed hummock when it has a bulged geometry (Figs. 12.6A, 12.6B, 12.6C, 12.6D, and 12.6E).

Fig. 12.6A Sorted circles surrounded by doughnut-like rounded ridges developed on a raised marine terrace, northern coast of Hornsund, southern Spitsbergen. The rod is 1 m long (Published with permission of © Zbigniew Zwoliński 2015. All Rights Reserved)

Fig. 12.6B Sorted polygons, Trail Ridge, Rocky Mountain National Park, Colorado (Published with permission of © Francisco Gutiérrez 2015. All Rights Reserved)

Fig. 12.6C Sorted stripes, Trail Ridge area, Rocky Mountain National Park, Colorado (Published with permission of © Francisco Gutiérrez 2015. All Rights Reserved)

Fig. 12.6D Non-sorted stripes south of Martin Lake, Cordova area, coastal Alaska. Antislope scarp related to flexural-slip faulting in the lower part of the slope (Published with permission of © Francisco Gutiérrez 2015. All Rights Reserved)

Fig. 12.6E Sorted steps developed on slope deposits, western slopes of Løvehovden, central Spitsbergen (Published with permission of © Janina Borysiak 2015. All Rights Reserved)

12.7 Pingo

Pingo is an Inuktitut word used in the geomorphological literature to designate mounds or hills related to the growth of a laccolith-like ice body in the subsurface and the upwarping of the overlying unconsolidated sediments. These ice-cored mounds, also known as hydrolaccoliths, can reach as much as 50 m in height and 600 m in basal diameter, and the ice volume can exceed 1 million m^3. They are found in large number mainly in Arctic areas. Two mechanisms related to contrasting groundwater conditions have been proposed to explain the growth of the ice core associated with continuous or discontinuous permafrost. Hydraulic or open-system pingos form by localised upward discharge of groundwater under hydraulic pressure. Hydrostatic or closed-system pingos are related to the drainage of a lake and the aggradation of ice by the migration of interstitial water from the unfrozen lake deposits. The growth of the ice core causes the thinning and stretching of the cover, resulting in the development of radial tension cracks. The ice core may be exposed by the breakage of the thermally insulating overburden in the crestal zone, leading to the development of a central crater. As the ice core melts, concentric cracks and inward-facing scarps develop, ultimately leading to the development of an enclosed depression with a lake and surrounded by an annular ridge. Available data on growth rates indicate that the age of some present-day pingos may be of the order of thousands of years (Fig. 12.7).

12.8 Palsa

Palsa is a Fennoscandian term that describes mounds or ridges of peat protruding from flat boggy areas within the discontinuous permafrost zone. In contrast to pingos, they have a permanent frozen core of peat and/or fine-grained sediment with lenses of segregated ice, rather than massive ice. Mound-like palsas may reach 150 m across and more than 10 m in height. Ridge-like palsas may be as much as 500 m long. The formation and preservation of the frozen core is related to the contrasting thermal conductivity of the wet or dry overlying peat. In autumn, the relatively high thermal conductivity of the water-saturated peat favours the freezing of the underlying sediment and the formation of segregated ground ice. In summer, the dry peat behaves as an insulating cover preventing the thawing of the core. The distribution of palsas seems to be related to deeper winter frost penetration in areas with thin snow cover. Once the ground surface starts to uplift, the wind preferentially removes the snow from those areas further enhancing the freezing process. Palsas typically show extensional cracks related to the upwarping of the deposits, may form complexes and are frequently associated with ponded thermokarst depressions (Fig. 12.8).

12.9 Alas

Alas is a local word from Yakutia, Siberia, that refers to thermokarst subsidence depressions characterised by steep sides and a flat floor. These enclosed basins with circular or elongated geometry may be occupied by lakes and reach more than 10 km in length and as much as 40 m in depth. Alas depressions are typically found in flat taiga environments and may form densely packed complexes that riddle extensive areas. The initiation of these thermokarst depressions requires a natural or anthropogenic disturbance in the thermal equilibrium of the ground (e.g. climate change, forest fire, deforestation). The thawing of ground ice leads to differential ground subsidence and the development of an irregular topography. Surface water accumulated in the depressions further accelerates the degradation of the permafrost (talik expansion) with the accompanying deepening and enlargement of the depressions. Adjacent alases may coalesce to form long depressions called alas valleys, although the basins do not result from erosion but from thermal-induced near-surface deformation processes (Figs. 12.9A and 12.9B).

Fig. 12.7 Ice-cored pingo in Yakutia, Siberia (Published with permission of © Cezary Kabała 2015. All Rights Reserved)

Fig. 12.8 Palsas around 2 m high in western Utsjoki, Finnish Lapland. The dark brown wind-abraded surfaces are underlain by peat uncovered by vegetation (Published with permission of © Matti Seppala 2015. All Rights Reserved)

Fig. 12.9A Alas depressions on a terrace of the Lena River, Yakutia, Siberia. Note the meandering rivers with abandoned oxbow lakes on the left side (Image from ©Bing Maps)

Fig. 12.9B Water-logged alas depressions, Yakutia, Siberia (Published with permission of © Cezary Kabała 2015. All Rights Reserved)

13 Desert and Aeolian Landforms

Desert environments are characterised by limited precipitation and sparse or non-existent vegetation cover. Deserts can be defined and classified according to the degree of aridity, as a measure of moisture availability. The aridity index, given by the ratio between precipitation and potential evapotranspiration, is used to differentiate between hyperarid (<0.05), arid (0.05–0.2) and semiarid regions (0.2–0.5). According to this criterion, around 37 % of the Earth's continental surface has arid conditions comprising 18 % semiarid, 12 % arid and 7 % hyperarid. A simpler classification is based on annual average precipitation: semiarid (200–500 mm), arid (25–250 mm) and hyperarid regions, where periods longer than twelve months without any precipitation are not infrequent. The main factors that control aridity in deserts include the following: (1) a zonal factor associated with the distribution of stable high-pressure cells in the subtropics (e.g. Sahara and Arabian deserts); (2) the continentality, based on the distance to the oceans that constitute the main source of moisture (e.g. deserts in Central Asia); (3) orographic barriers that block the path of weather fronts creating rain shadows (e.g. desert regions east of Sierra Nevada in the USA); and (4) cold oceanic currents that restrict evaporation and atmospheric moisture (e.g. the Namib-Angola desert).

The extreme environmental conditions of desert areas have a strong impact on the morphogenetic processes and determine the development of specific landforms. The scant precipitation has a significant impact in most geomorphic systems including hillslopes, rivers or lakes, as well as on weathering and pedogenic processes. The limited water availability results in bare surfaces or with a patchy vegetation cover, which is the main requisite for aeolian activity. Slopes, with a reduced protective plant cover, are highly sensitive to extreme events and environmental variations. Relatively small changes in the vegetation cover may lead to changes in the behaviour of the slopes, from aggradational to incisional or vice versa. Run-off in desert areas tends to be enhanced by limited rainfall interception and infiltration due to the sparse vegetation and the presence of duricrusts, respectively. Channel flow in deserts is mostly ephemeral, with the exception of allochthonous rivers that drain water from nondesert areas (e.g. Colorado River). The flow events are frequently related to high-intensity and short-lasting rainstorm events, which may generate sudden sediment-laden flash floods; these have a large geomorphic effectiveness and a high damaging potential. Channels may be discontinuous due to water loss by infiltration or can be blocked by dune fields. They frequently flow into the bottom of endorheic basins, where playa lakes develop. These are lakes that flood sporadically and are characterised by evaporitic deposition and extremely flat floors controlled by the water table, which determines the deepest level for aeolian deflation.

Wind activity creates erosional and depositional landforms that cover approximately 20 % of the Earth's land surface. The effectiveness of the wind as a geomorphic agent largely depends on its energy, the lack of vegetation and the availability of lose particles, especially sand. Erosional landforms may be generated by different mechanisms. Deflation is the removal of loose material and generates desert pavements, pans and blowouts. Abrasion involves the wearing down of cohesive material by the impact of wind-transported particles. It essentially operates close to the surface and produces ventifacts or streamlined ridges called yardangs. Aeolian deposition leads to two main types of accumulations, sand dunes with a wide variety of morphologies and deposits of windblown dust or loess that tend to drape the land surface and may reach hundreds of metres thick. Aeolian dunes are bedforms mostly consisting of wind-transported sand. They may cover extensive tracts of land called ergs (sand seas) or dune fields, depending on whether they are larger or smaller than 100 km^2, respectively. The sand seas in the Arabian Peninsula cover around 770,000 km^2, notably the Rub' al Khali extending 560,000 km^2. The main sources of sand include beaches, dry playa lakes, fluvial floodplains and deltas. Dune morphology largely depends on the rate of

F. Gutiérrez, M. Gutiérrez, *Landforms of the Earth*, DOI 10.1007/978-3-319-26947-4_13

sand supply, the directional variability of the wind and the presence of obstacles that may control the development of anchored dunes.

A great part of the environmental problems in arid areas are directly or indirectly related to the limited water availability. These include hazardous processes like salt weathering, dust storms, dune encroachment or flash floods in ephemeral streams that are commonly perceived as harmless. Groundwater pumping in numerous arid regions is causing the exhaustion of a non-renewable resource, ground subsidence and the development of earth fissures. The abstraction of water from rivers is rapidly altering the dynamics of the fluvial systems and causing rapid declines in the level of the lakes they feed (e.g. Dead Sea, Aral Sea). In addition, irrigation in arid areas may result in the salinisation of the soils by the accumulation of salts, favoured by the high evapotranspiration rates that prevent their lixiviation.

13.1 Playa Lake

Playa is a Spanish word widely used in the literature to designate temporary lakes that flood periodically. They are also designated with many other terms like salt flat, alkali flat, salinas, salada, salar and chott. These ephemeral lakes mostly occur in the lowermost sectors of endorheic basins where groundwater discharges. They are characterised by a negative net hydrologic budget; the evaporation potential overwhelms water inflow by groundwater discharge, run-off and precipitation. Consequently, most playas occur in arid and semiarid areas. Deposition in playas is dominated by fine-grained detrital facies (clay and silt) and evaporites that may precipitate within the sediments, on the lake bottom or in the water column. They may be found in a wide variety of settings such as active tectonic basins, deflation hollows, abandoned or blocked drainages, karst depressions or impact and volcanic craters. These lakes are characterised by a very flat topography controlled by the water table and the capillary fringe. When the lake is dry, the lowering of the lake bottom by aeolian deflation is stopped once the surface meets cohesive deposits associated with the water table. The level topography implies that relatively small inputs and outputs of water involve large expansions and contractions of the flooded area, respectively (Figs. 13.1A and 13.1B).

13.2 Desiccation Cracks

Desiccation cracks develop in unconsolidated, fine-grained, cohesive deposits. They are very common in playas and mud flats. Evaporation from these water-saturated deposits results in a volume reduction, which is accommodated by vertical contraction (subtle subsidence) and the development of cells affected by inward horizontal contraction. The latter tensional stresses generate cracks with a polygonal pattern (Fig. 13.2A). The attributes of the cracks show a great variability that mainly depends on the characteristics of the deposit and the evolutionary stage of the process. The density and width of cracks are commonly higher in finer-grained deposits, especially if they include swelling clay minerals like montmorillonite. The surface of the intervening blocks may be flat, convex (commonly related to salt precipitation) or concave (indicative of rapid evaporation). The latter may show thin lamina of clay rolled inwards called mud curls (Fig. 13.2B). Raindrop impressions are also common on these surfaces. Large polygonal cracks or fissures may also develop by contraction related to dewatering of unconsolidated sediments by groundwater withdrawal. These large fissures, which may enlarge rapidly by water erosion, can cause significant damage to human structures (Fig. 13.2C).

Fig. 13.1A Playa lake with a saline crust north of Antero Reservoir, South Park tectonic basin, southern Rocky Mountains, Colorado. The playa occurs in the bottom of a subsidence depression related to deep-seated dissolution of evaporites superimposed on a Middle Pleistocene outwash plain (Published with permission of © Francisco Gutiérrez 2015. All Rights Reserved)

Fig. 13.1B Playas developed on a structural platform mainly underlain by gypsum and limestone bedrock. The origin of these lake basins with yardangs in the downwind side (foreground) is related to the combined effect of dissolution and aeolian deflation. The lake in the background, called La Playa, probably inspired the Spaniards who colonised the American Southwest and gave this name to the salt lakes of that region (Published with permission of © Francisco Gutiérrez 2015. All Rights Reserved)

Fig. 13.2A Desiccation cracks developed on clayey deposits in the bottom of a playa in Dobi Graben, Ethiopia (Published with permission of © Francisco Gutiérrez 2015. All Rights Reserved)

Fig. 13.2B Desiccation cracks with curved lamina on overbank fines in the Draa Valley, Morocco. Erg Chigaga in the background (Published with permission of © Francisco Gutiérrez 2015. All Rights Reserved)

Fig. 13.2C Large cracks with an overall polygonal pattern developed on a mantled sandy-gravelly pediment in Bayda, west of Hail, Saudi Arabia. These fissures, around 1 m deep as revealed by trenching, seem to be related to groundwater withdrawal (Published with permission of © Francisco Gutiérrez 2015. All Rights Reserved)

13.3 Salt Polygons

Salt polygons are a type of patterned ground that commonly forms in the floor of playas and sabkhas with saline waters. Evaporation leads to the formation of a salt crust associated with desiccation cracks. Salt minerals precipitate sequentially following an ascending solubility order; commonly carbonates, sulphates and chlorides. As precipitation proceeds favoured by capillary pumping of brines, the salt plates thicken and grow laterally over the cracks due to higher evaporation in the flanks. The expansion of the crust may produce buckle structures, called tepees, and thrusts, commonly related to more rapid precipitation in the flanks facing upwind (Figs. 13.3A and 13.3B). Additionally, extrusion of mud or preferred salt precipitation at the plate edges may lead to the formation of mud or salt ridges, respectively (Fig. 13.3C). Salt blisters are also common on these ephemeral polygonal surfaces, which are rapidly obliterated by water inflow events.

13.4 Desert Varnish

Desert varnish is a thin and dark deposit, typically less than 0.1 mm thick, that coats bedrock surfaces and clasts in arid areas. It consists of clay minerals (more than 50 %) deposited as dust and cemented by manganese and iron hydroxides and oxides. Its formation requires exposed rock surfaces that remain stable for thousands of years. The manganese gives a dark brown to black colour, whereas reddish tones are linked to iron. It has a controversial origin that seems to be related to the combination of inorganic chemical processes and bacterial activity responsible for the enrichment of manganese and iron (Figs. 13.4A and 13.4B).

13.5 Calcrete

Calcretes are near-surface accumulations of secondary calcium carbonate, also known as caliche, carbonate crust and petrocalcic horizon. It is essentially a white and cream-coloured pedogenic mineral deposit that may be derived from the parent material, airborne dust and rainfall. The calcium and bicarbonate ions are leached by vadose downward flow and accumulate preferentially in the B illuviation soil horizon. Calcretes evolve through multiple morphological stages of calcium carbonate accumulation that may be used, together with the amount of secondary carbonate, as tools for relative dating and correlation. In gravelly deposits, initially, the clasts are coated by discontinuous and then continuous carbonate. In a subsequent stage, the secondary carbonate forms massive accumulations between clasts producing a

Fig. 13.3A Salt crust with metre-scale polygons displaying upwind-facing overthrust edges (Salada de Mediana, Ebro Cenozoic Basin, NE Spain) (Published with permission of © Francisco Gutiérrez 2015. All Rights Reserved)

Fig. 13.3B Salt polygons with tepees and thrusts along their edges (Dead Sea, Israel) (Published with permission of © Francisco Gutiérrez 2015. All Rights Reserved)

plugged hardpan. Then, calcium carbonate accumulates as laminae on the upper part of the hardpan. The more evolved stages are characterised by multiple generations of brecciation, pisolith formation and accumulation of lamina around the blocks. Calcretes developed on fine-grained deposits grade from a powdery to a nodular texture until the hardpan is developed. Deposits cemented by caliche may behave as resistant protective layers similarly to caprocks, favouring the preservation potential of deposits and geomorphic surfaces, as well as relief inversion; a surface once formed on lowlands forms the top of prominent reliefs (Figs. 13.5A and 13.5B).

13.6 Talus Flatirons and Relict Slope Rings

Talus flatirons, also known as triangular slope facets and tripartite slopes, are relict slopes detached from the source that once supplied the debris that armour their surface. These landforms are commonly associated with mesas, buttes and cuestas, where a resistant scarp-forming caprock overlies easily erodible sediments. They typically have a triangular or trapezoidal geometry with the apex pointing towards the scarp. The development of talus flatirons involves alternating stages of accumulation and incision in the slopes. In the initial stage, detrital material derived from the erosion of the rock scarp accumulates on the debris slope. Subsequent dissection (rilling and gullying) of the debris slope accompanied by the retreat of the free face scarp leads to the disconnection of the talus flatirons from their sediment source. Multiple alternating periods of accumulation and incision may result in the generation of talus flatiron chronosequences with paleoenvironmental meaning (Fig. 13.6A). In small buttes, the complete removal of the caprock may lead to a process of relief inversion. The soft bedrock in the core of the butte may be eroded and evacuated at a much faster rate than the more resistant colluvium armouring the slopes, resulting in the formation of relict slope rings (Fig. 13.6B).

13.7 Pediment

Pediments are gently inclined surfaces cut across bedrock that connect uplands dominated by erosion with the local base level (e.g. river, lake). They display a concave longitudinal profile, and their gradient is generally lower than 6°. Pediments commonly lack any cover in the proximal sector,

Fig. 13.3C Salt polygons with acicular crystals, extrusions of mud and preferred salt precipitation along the edges (Badwater Lake, Death Valley National Park, California) (Published with permission of © Francisco Gutiérrez 2015. All Rights Reserved)

whereas in the distal sectors pediments may have a thin veneer of detrital deposits. The formation of pediments (pedimentation) is attributed to the progressive retreat of the slopes in the upland areas by a general process of backwearing. Consequently, they are time-transgressive erosional surfaces progressively younger towards the piedmont angle. Extensive erosional surfaces resulting from the expansion and coalescence of pediments are called pediplains, which occur in regions with long-term tectonic stability. Several hypotheses have been proposed to explain the development of these probably polygenetic landforms including surface and subsurface weathering and fluvial processes dominated by unconfined flows (sheetflooding) (Fig. 13.7).

13.8 Desert Pavement

Desert pavements, also named regs, are bare and markedly level surfaces covered by clasts that commonly armour an underlying bed of fine-grained deposits. The clasts may be polished, faceted by aeolian abrasion and coated by desert varnish. They usually develop on relict alluvial deposits or weathered bedrock. The most widely accepted genetic interpretation is the removal of the fine surface material by deflation and/or overland flow, leaving behind a concentration of gravelly particles that cannot be entrained by air currents. Equivalent stone pavements can be found in periglacial environments (Fig. 13.8).

Fig. 13.4A Sandstone largely coated by a dark desert varnish that contrasts with the lighter-coloured fresh exposures affected by weathering (Canyonlands National Park, Utah) (Published with permission of © Francisco Gutiérrez 2015. All Rights Reserved)

13.9 Ventifact

Ventifacts are gravels and boulders faceted by wind abrasion. The exposed surfaces are reshaped to form flat sides that intersect along sharp edges. Abrasion is caused by wind-blown sand and dust particles transported by saltation and suspension and may affect the lee side of the clasts by vorticity. Facets show an overall polished appearance and display small-scale erosional features like pits or grooves. Ventifacts with one or three facets are called einkanter and dreikanter, respectively. They are common in desert pavements and may be also found in periglacial and coastal environments. The facets of relict ventifacts with pervasive orientation may be used to infer paleowind directions (Fig. 13.9).

13.10 Yardang

Yardang designates ridges carved by wind erosion either on bedrock or cohesive surficial sediments. These erosional features may form extensive swarms with intervening corridors showing a striking parallelism controlled by the regional wind trend. The overall morphology of most yardangs resembles an upturned ship's prow, with a steep and frequently undercut upwind front, and a more gentle downwind sector. Yardangs show a wide range of dimensions, from a few centimetres to tens of metres high and several kilometres long. Their formation is mainly related to wind erosion processes (abrasion, deflation) in combination with fluvial action and mass movements. They mostly occur in arid and hyperarid areas, where they may occupy vast expanses of the order of hundreds of thousands square kilometres (Tibesti in the Sahara Desert, Lut Desert in Iran), but they have been also reported in semiarid areas. Yardangs carved in soft sediments like paleolake deposits may form in a relatively short time span (Figs. 13.10A and 13.10B).

13.11 Pan and Lunette

Lunettes are ridge-like transverse dunes that form along the downwind margin of shallow and flat-floored playa lakes (also called pans). The use of this French word is related to the common crescentic shape of these dunes, controlled by the curved edges of the adjacent pans. The pan and lunette association strongly suggests that deflation plays a major

Fig. 13.4B Desert varnish on a sandstone cliff with petroglyphs in Wadi Rum, Jordan (Published with permission of © Francisco Gutiérrez 2015. All Rights Reserved)

role in the development of the former. In fact, the depressions tend to be oriented with respect to the regional winds, and the lee sides commonly have an arcuate trace with the convexity pointing downwind. A peculiarity of lunette dunes is that they may consist of a large proportion of clay (clay dune), in contrast with common dunes that are mainly composed of sand. The clay is derived from the floors of the adjacent playas during drying periods as sand-sized peloids of aggregated clay, frequently cemented with salts. Wetting and drying cycles, salt weathering and other biogenic processes favour the formation of loose clay aggregates in the pan floors. Lunettes are commonly a few metres to a few tens of metres high and may reach some kilometres in length. Their sides generally have lower gradients than common sand dunes, with a steeper upwind side. Some pans display multiple generations of lunettes separated by swales that may constitute a good source of paleoenvironmental information (Figs. 13.11A and 13.11B).

13.12 Barkhan Dune

Barkhans are isolated crescent-shaped sand dunes with a steep concave downwind face and a gentler windward side. The sharp crest is oriented perpendicular to the wind direction in the central zone and curves downwind in the flanks (Fig. 13.12A). These dunes develop on flat firm surfaces in areas with limited sand supply and unidirectional winds. Barkhans have commonly a metre-scale height, although dunes as high as 100 m have been reported. These dunes may

Fig. 13.5A Caliche with massive accumulation of secondary carbonate (plugged hardpan) developed on a pediment deposit mostly consisting of basaltic gravel (Playa de Sotavento, Fuerteventura, Canary Islands) (Published with permission of © Francisco Gutiérrez 2015. All Rights Reserved)

Fig. 13.5B Polished sample of a calcrete developed on a gravelly mantled pediment deposit showing an advanced stage of development (Ebro Cenozoic Basin, NE Spain). Massive accumulation of carbonate between clasts has resulted in the formation of a plugged hardpan. Subsequent brecciation favoured formation of laminar carbonate in the lower side of the blocks (Published with permission of © Francisco Gutiérrez 2015. All Rights Reserved)

Fig. 13.6A Multiple generations of talus flatirons developed around a mesa with a caprock of indurated chert of the Cretaceous Mishash Formation, underlain by soft chalky sediments (Neve Zohar area, west margin of the Dead Sea Basin, Israel) (Published with permission of © Francisco Gutiérrez 2015. All Rights Reserved)

Fig. 13.6B Relict slope ring resulting from relief inversion in a butte after the removal of the caprock (Larkspur, Colorado Piedmont) (Published with permission of © Francisco Gutiérrez 2015. All Rights Reserved)

Fig. 13.7 Pediment cut across igneous rocks of the Arabian Shield and locally interrupted by residuals. The planar and gently sloping surface is extensively covered by a thin veneer of aeolian sand (Hail area, Saudi Arabia) (Published with permission of © Francisco Gutiérrez 2015. All Rights Reserved)

Fig. 13.8 Desert pavement in Bsita, Al Jouf region, Saudi Arabia. The pavement shows a high concentration of chert clasts, less vulnerable to weathering than the accompanying limestone and sandstone (Published with permission of © Francisco Gutiérrez 2015. All Rights Reserved)

Fig. 13.9 Ventifact with desert varnish and three polished facets intersecting at sharp angles (dreikanter) in Death Valley National Park, California (Published with permission of © Francisco Gutiérrez 2015. All Rights Reserved)

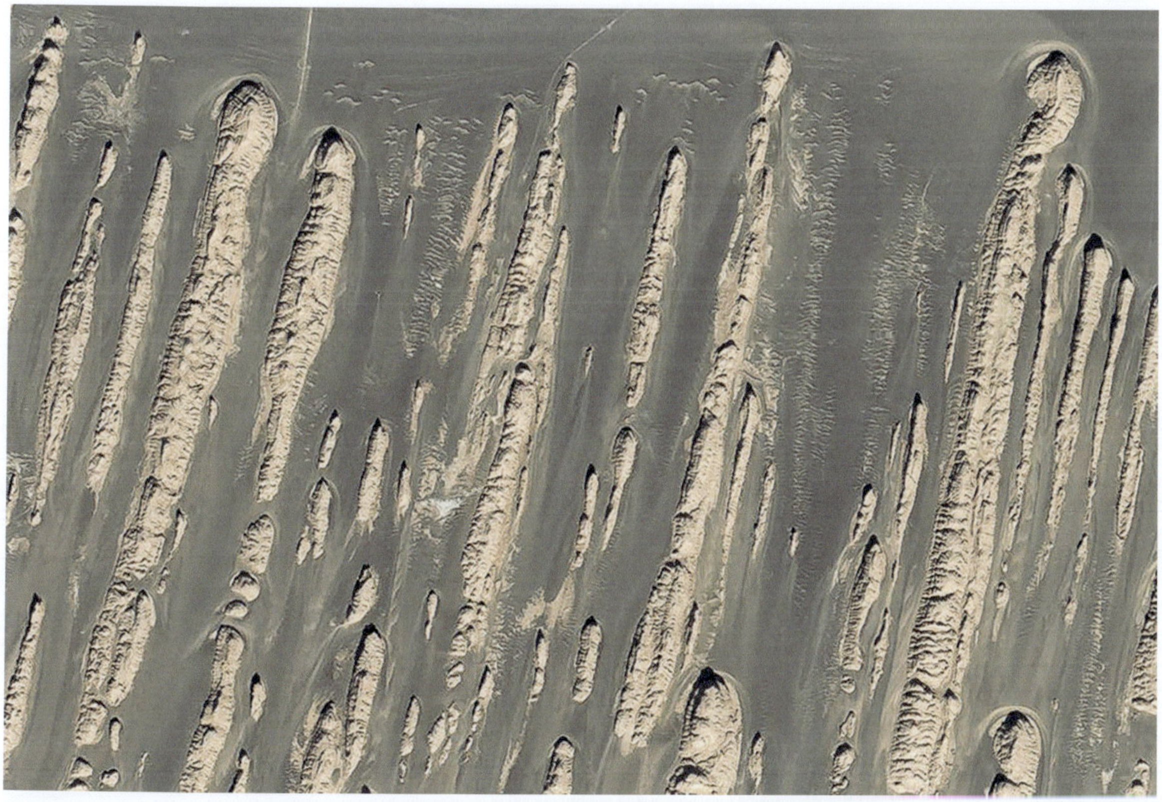

Fig. 13.10A Swarm of wind-fluted yardangs in Xinjiang area, Gobi Desert, China (Image from © Bing Maps)

Fig. 13.10B Yardangs carved in subhorizontal gypsum bedrock by a strong wind called cierzo in the semiarid central sector of the Ebro Cenozoic Basin, Quinto area, NE Spain (Published with permission of © Francisco Gutiérrez 2015. All Rights Reserved)

Fig. 13.11A Lake Sunday, a pan oriented perpendicularly to the dominant wind with lunette dunes along the curved downwind margin (Yorketown, South Australia) (Image from © Bing Maps)

Fig. 13.11B Guallar playa with a curved lunette dune on the downwind edge, at the *left* (Bujaraloz, Ebro Cenozoic Basin, NE Spain). The depression is oriented in the direction of the prevailing wind (cierzo) and has a wider downwind half and a pointed upwind edge (Published with permission of © Francisco Gutiérrez 2015. All Rights Reserved)

Fig. 13.12A Barkhan dunes or reg in the Draa Valley, Morocco (Published with permission of © Francisco Gutiérrez 2015. All Rights Reserved)

migrate at high velocities, which are inversely proportional to their heights. As sand supply increases, barkhans merge laterally to form barkhanoid ridges (Fig. 13.12B).

13.13 Transverse Dune

These are sand ridges perpendicular to the dominant wind direction with asymmetric profile, consisting of a steeper concave downwind slope and a more gentle upwind side. The crests of transverse dunes are typically sharp and slightly sinuous and tend to display regular spacing. Transverse dunes commonly develop in vegetation-free areas with unimodal wind regime and higher sand supply than the barkhan areas (Fig. 13.13).

13.14 Parabolic Dune

These are U- or V-shaped dunes in plan view with a convex active nose pointing downwind and two frequently vegetated arms (Fig. 13.14). The arms or horns, in contrast to barkhans, point upwind. The crest of parabolic dunes may reach tens of metres in the central sector and the arms attain kilometre-scale lengths. This dune type occurs in areas with unidirectional wind regime and sparse vegetation cover. It is particularly common in dune fields developed in coastal areas and semiarid areas. Parabolic dunes may migrate at relatively high rates, inversely proportional to their heights. Parabolic dunes and deflation basins like pans or blowouts may have a joint origin. Deflation of loose sand or other particles from the core of the dune may create such depressions (Fig. 13.14).

13.15 Longitudinal Dune

Longitudinal dunes, also known as linear or seif dunes, form swarms of approximately parallel and regularly spaced sand ridges that may reach 20 km in length and 200 m in height. The sharp or rounded crestlines and the intervening troughs may be straight or sinuous. Longitudinal dunes commonly show a composite cross profile, with lower gentle slopes, often vegetated, and upper abrupt slopes with greater aeolian activity. The net sand transport is parallel to the crest. Longitudinal dunes develop in areas with high sand supply and bimodal winds or with a broad unidirectional regime. They are the most common dune type in the great sand deserts worldwide (Fig. 13.15).

Fig. 13.12B Barkhanoids resulting from the lateral coalescence of crescentic dunes in Al Jouf area, Saudi Arabia. The steep concave side of the dunes indicates a prevailing wind direction to the left (Published with permission of © Francisco Gutiérrez 2015. All Rights Reserved)

Fig. 13.13 Transverse dunes in Al Jouf area, Saudi Arabia. Barkhanoids in the background. Dominant wind is to the left (Published with permission of © Francisco Gutiérrez 2015. All Rights Reserved)

Fig. 13.14 Parabolic dunes in White Sands National Park, Tularosa Basin, Alamogordo, New Mexico. These dunes composed of gypsum particles are located in the marginal zone of an erg developed in the downwind side of Lucero Lake, a playa with gypsum deposition. Wind direction is towards the upper-right corner (Image from © Google Earth)

Fig. 13.15 Satellite image of longitudinal dunes in the southwestern sector of the Rub' al Khali desert (Empty Quarter), Yemen. The dune ridges have a spacing of around 2–4 km (Image from © Bing Maps)

13.16 Star Dune

Star dunes, also designated as ghourds, are characterised by a pyramidal geometry with several sharp ridges that radiate from a central peak or crest. These are the largest dunes in most ergs, with heights of up to 400 m. The slopes show a general concave profile, with steeper gradients in the upper part. They commonly occur as evenly spaced or coalesced groups and may be superimposed by other dune types forming compound dunes or draa. Star dunes develop in areas with multidirectional wind regimes. They also occur associated with topographic barriers that contribute to increase the directional variability of the wind. The sand is affected by significant seasonal reworking, but the overall migration rate of the dunes is very low (Figs. 13.16A, 13.16B, 13.16C, and 13.16D).

13.17 Climbing Dune

These are topographically controlled dunes that occur on windward-facing slopes, commonly steeper than 30°. The sand is trapped by the obstacle, and the accumulation grows vertically and upslope forming a climbing dune or a sand ramp. When the wind-transported sand surmounts the hill and accumulates on the downwind side of the topographic obstacle, it forms a falling dune. Climbing dunes are common in desert areas, are typically isolated and may reach more than 100 m in height (Fig. 13.17).

13.18 Nebkha Dune

These are small dunes anchored by plants. The sand is deposited on the trapping plant and along its lee side, forming an elongated dune parallel to the wind direction (Fig. 13.18). The dune height may reach around 3 m and decreases away from the plant. They are also known as shadow dunes, coppice dunes or shrub dunes, among other terms. These dunes are common in semiarid areas, in the marginal sectors of dune fields and sand seas and in coastal zones. The deposition of sand on the plant and its partial burial may stimulate the upward growth of the plant. Consequently, the vegetation traps additional sand at progressively higher elevation, ultimately resulting in the built-up of vegetated mounds up to several metres high.

Fig. 13.16A Satellite image of an isolated star dune in the eastern sector of the Rub' al Khali (Empty Quarter), Saudi Arabia (Image from © Bing Maps)

Fig. 13.16B Star dune in the Gobi Desert, Dunhuang, Gansu Province (Published with permission of © Scott Weatherson 2015. All Rights Reserved)

Fig. 13.16C Star dune in Erg Chigaga, Draa Valley, Morocco (Published with permission of)

Fig. 13.16D Star dunes and pans in Sossusvlei, Namib Desert, Namibia (Published with permission of)

Fig. 13.17 Climbing dune more than 60 m high developed on the windward steep slope of a hill west of Hail, Saudi Arabia (Published with permission of © Francisco Gutiérrez 2015. All Rights Reserved)

Fig. 13.18 Nebkhas associated with plants in the coastal strip south of Walvis Bay, Namib Desert, Namibia (Published with permission of © Francisco Gutiérrez 2015. All Rights Reserved)

13.19 Blowout

Blowouts are enclosed hollows excavated by wind erosion (deflation) into loose aeolian sand. They may have a saucer- or trough-like morphology and the long axis is commonly parallel to the direction of the wind. Blowouts are particularly common in partially stabilised dune fields of coastal areas, but may also develop in other environments and in vegetation-free active dunes. They tend to occur in the more elevated areas of vegetated dune landscapes and may be induced by the local removal of the protective vegetation. The wind is funnelled into the depressions, where there is an increase in flow velocity and vertical vortexes may develop. Blowouts tend to be ephemeral landforms affected by rapid changes and with a limited persistence time. Accumulation of sand in the downwind side may lead to the development of parabolic dunes (Figs. 13.19A and 13.19B).

Fig. 13.19A Steep-sided blowout excavated in active dunes in Bolonia Bay, Atlantic coast of southern Spain. The elongated hollow is oriented in the direction of the prevailing wind, to the *left* of the image (Published with permission of © Francisco Gutiérrez 2015. All Rights Reserved)

Fig. 13.19B Blowout in Al Liwa, Rub' al Khali, United Arab Emirates (Published with permission of © Mateo Gutiérrez 2015. All Rights Reserved)

Further Reading

Dictionaries and Encyclopedias

Fairbridge RW (ed) (1968) The Encyclopedia of Geomorphology. Dowden, Hutchinson & Ross. Stroudsburg 1295 p.

Goudie AS (ed) (2004) Encyclopedia of Geomorphology. Routledge 2 volumes.

Goudie A, Atkinson, BW, Gregory, KJ, Simmons IG, Stoddart DR, Sugden D (eds) (2000) The Encyclopedic Dictionary of Physical Geography. Blackwell. Oxford 611 p.

Shroder JF (ed) (2013) Treatise on Geomorphology. Elsevier-Academic Press. Amsterdam 14 vols.

Whittow JB (1984) Dictionary of Physical Geography. Penguin. London 591 p.

Geomorphology Textbooks

Anderson RS, Anderson SP (2010) Geomorphology. The Mechanics and Chemistry of Landscapes. Cambridge University Press. Cambridge 637 p

Chorley RJ, Schumm SA, Sugden DE (1984) Geomorphology. Methuen. London 605 p.

Gregory KJ, Goudie AS (eds) (2011) The SAGE Handbook of Geomorphology. Sage. Los Angeles 610 p.

Gutiérrez M (2013) Geomorphology. CRC Press/Balkema. Leiden 1017 p.

Harvey A (2012) Introducing Geomorphology. A Guide to Landforms and Processes. Dunedin. Edinburgh 124 p.

Hugget RJ (2003) Fundamentals of Geomorphology. Routledge. London 386 p.

Structural and Tectonic Geomorphology

Burbank DW, Anderson RS (2012) Tectonic Geomorphology. Wiley-Blackwell. Chichester 454 p.

Keller EA, Pinter N (1996) Active Tectonics. Earthquakes, Uplift and Landscape. Prentice Hall. New Jersey 338 p.

McCalpin JP (ed) (2009) Paleoseismology. Elsevier. Amsterdam 613 p.

Ollier CD (1981) Tectonics and Landforms. Longman. Essex 324 p.

Volcanic Geomorphology

Schmincke H-U (2004) Volcanism. Springer. Berlin 324 p.

Sigurdsson H (ed) (2000) Encyclopedia of Volcanoes. Academic Press. San Diego 1417 p.

Karst Geomorphology

Ford D, Williams P (2007) Karst Hydrogeology and Geomorphology. John Wiley & Sons. Chichester 562 p.

Gunn J (ed) (2004) Encyclopedia of Caves and Karst Science. Fitzroy Dearborn. New York 902 p.

Palmer AN (2007) Cave Geology. Allen Press. Lawrence, Kansas 454 p.

Sweeting MM (1972) Karst Landforms. Macmillan. London 362 p.

Granite Geomorphology

Twidale CR, Vidal-Romaní JR (2005) Landforms and Geology of Granite Terrains. Balkema. Leiden 351 p.

Migoń P (2006) Granite Landscapes of the World. Springer, Oxford 384 p.

Weathering

Bland W, Rolls D (1998) Weathering. Arnold. London 271 p.

Ollier CD (1984) Weathering. Longman. London 270 p.

Slope Movements

Crozier MJ (1986) Landslides. Causes, Consequences and Environment. Croom Helm. London 252 p.

Cruden DM, Varnes DJ (1996) Landslide Types and Processes. In: Schuster RL (ed) Landslides. Investigation and Mitigation. National Academy Press. Washington, pp 36–75.

Dikau R, Brunsden D, Schrott L, Ibsen M-L (eds) (1996) Landslide Recognition, Identification, Movement and Causes. John Wiley & Sons. Chichester, 251 p.

Fluvial Geomorphology

Charlton R (2009) Fundamentals of Fluvial Geomorphology. Routledge. London 234 p.

Knighton D (1998) Fluvial Forms and Processes. A New Perspective. Arnold. London 383 p.

Schumm SA (1977) The Fluvial System. John Wiley & Sons. Chichester 338 p.

F. Gutiérrez, M. Gutiérrez, *Landforms of the Earth*, DOI 10.1007/978-3-319-26947-4

Coastal Geomorphology

Bird E (2008) Coastal Geomorphology. An Introduction. John Wiley & Sons. Chichester 411 p.
Schwartz ML (ed) (2005) Encyclopedia of Coastal Science. Springer, Dordrecht 1211 p.
Trenhaile AS (1987) The Geomorphology of Rock Coasts. Clarendon Press. Oxford 384 p.

Glacial Geomorphology

Bennet MR, Glasser NF (1996) Glacial Geology. Ice Sheets and Landforms. John Wiley & Sons. Chichester 364 p.
Benn DI, Evans DJA (1998) Glacier & Glaciation. Arnold. London 734 p.

Periglacial Geomorphology

French HM (2007) The Periglacial Environment. John Wiley & Sons. Chichester 458 p.
Washburn A (1979) Geocryology. A survey of Periglacial Processes and Environments. Arnold. London 406 p.

Desert and Eolian Landforms

Parsons AJ, Abrahams AD (eds) (2009) Geomorphology of Desert Environments. Springer. Dordrecht 831 p.
Lancaster N (1995) Geomorphology of Desert Dunes. Routledge. London 290 p.
Livingstone I, Warren A (1996) Aeolian Geomorphology. An Introduction. Longman. Essex 210 p.
Thomas DSG (ed) (2011) Arid Zone Geomorphology. Process, Form and Change in Drylands. Wiley-Blackwell. Chichester 624 p.

Index

F. Gutiérrez, M. Gutiérrez, *Landforms of the Earth*, DOI 10.1007/978-3-319-26947-4

MIX
Papier aus verantwortungsvollen Quellen
Paper from responsible sources
FSC® C105338

If you have any concerns about our products,
you can contact us on
ProductSafety@springernature.com

In case Publisher is established outside the EU,
the EU authorized representative is:
Springer Nature Customer Service Center GmbH
Europaplatz 3, 69115 Heidelberg, Germany

Printed by Libri Plureos GmbH
in Hamburg, Germany